Aquarium Management

Aquarium Management

Compiled and Edited
by
Amita Saxena
Associate Professor
College of Fisheries
GBPUAT, Pantnagar - 263165, India

2021
Daya Publishing House®
A Division of
Astral International Pvt. Ltd.
New Delhi – 110 002

First Published, 2003
Reprinted, 2021
ISBN: 9788170352846

Published by : **Daya Publishing House®**
A Division of
Astral International Pvt. Ltd.
– ISO 9001:2015 Certified Company –
4736/23, Ansari Road, Darya Ganj
New Delhi-110 002
Ph. 011-43549197, 23278134
E-mail: info@astralint.com
Website: www.astralint.com

Digitally Printed at : **Replika Press Pvt. Ltd.**

Contents

Preface

Aquarium and ornamental fisheries is a new and upcoming trade in India. This book provides basic knowledge of fish keeping in offices, hotels, tourists spots and of course in teaching. This is a comprehensive text to public regarding beautiful aquatic creatures, which are going towards extinction, need conservation. Through aquarium one can show the imitation of nature in a limited area and educate public for its conservation besides enhancing the beauty of place through living organisms.

Ornamental fish keeping is not only a hobby it involves a series of activities such as construction, and design of aquarium and ponds, breeding, feeding and diseases of ornamental fishes, have been discussed in this book. After knowing the details of Aquarium/ Ornamental fisheries one can fetch national and international market. Aquarium is a convenient tool for environmental and aquatic biology education and self employment. This book will be beneficial to fish hobbyist, scientists, naturalists, farmers, unemployed youth and students too.

I am thankful to my teachers, students, friends, brothers and sisters who helped me in doing such kind of work. I would like to thank the publisher Mr. Anil Mittal who has taken lots of pain to publish this book.

I welcome the constructive criticism from my learned readers specially students, scientists, farmers and general public to improve the text in next edition.

Amita Saxena

Chapter 1

Introduction

An aquarium is an enclosed volume of water containing a mixture of selected and unselected captive living aquatic organism. It is essentially unstable and to obtain stability must be carefully designed and managed.

When water is placed in aquarium tank, change in water chemistry begins and this accelerates as the stocking density of the tank is increased. In aquarium, the water is continuously recycled. Such systems are more independent of the environmental changes which effect open systems but the aquarium inhibits and to change the composition of the aquarium water.

When setting up an aquarium, large or small, the aim is to arrest the process of deterioration and thereafter exercise control over water quality.

The main features are to ensure a supply of good quality water, and then to maintain its quality by the removal of accumulated metabolites and by maintenance of the correct level of pH and dissolved gases.

It is a common fallacy that such a tank can consist of a balanced system of animals, plants and bacteria whose water quality control is effected by the living organisms themselves. If the aquarium tank does not contain plants, water quality may be improved by the application of a filter.

Aquarium

The most convenient type of aquarium is made from a welded frame of angle iron in which glass is set with strong sealing compound like bitumen. Such an aquarium is strong and yet relatively light in weight.

Saving can be effected by purchasing the angle iron and having this welded to form a frame of the selected size, and then putting in the glass and fixing it with battery sealing compound bitumen at home. Good workmanship is necessary on the frame with a trust will not rest firmly and will be difficult to glue. In general the width should be as great, or nearly as great, as the depth. Brass or Aluminium frame may be used. Decorative wooden frames can also be made from hard woods such as oak, and these frames are glazed just like the metal frame. They form highly decorative aquaria and are perfectly satisfactory if the points are strongly made with the metal frame. Special care should be taken with it before glazing to ensure good protection against rust formation. If the metal is thoroughly cleaned by brushing with a wire brush and made bright with a file and emery paper, one of the commercial rust-preventing preparations can be painted over two coats of aluminium paint may be applied – rusting is particularly liable to affect the inside top frame of the tank so this part should be given special attention. After antirusting and chromium coats an under coating of paint is given and then furnished off with a hard glass enamel paint of chosen colour.

Construction

There are several good kinds of 'cement' or 'putty' for aquarium use. These do not dry too hard, but form a cushion and water repellent layer between the glass and the frame, which ensures that the glass is not held so firmly at the risk of breakage occurs and yet prevents leaking by good adhesion to glass and to the frame surfaces. About 2½ lb of waterproof cement will be enough to glaze an aquarium of the dimensions recommended above i.e. 24" × 12" × 13".

The glass thickness required will depend on the size of the aquarium. Plate glass 8-10 mm thick is used for tanks up to about 4 ft in length and 15 inch deep. Aquaria less than 2 ft in length, and 15 inch in depth can be glazed with 4.2 mm glass. For the side, end and front glasses use clear glass but for the base added protection is given by using plate glass with a wire mesh in it. In estimating the sizes of the panes, remember that close contact should be made between them at the corners, and also make allowance for about 1/8 inch thickness of cement.

The base glass asbestos is always fitted first, so its dimensions need to be slightly less than those of the inside of the frame panels for the ends will be fitted in last, between the front and back panels, so their widths need to be reduced by an amount equal to the thickness of the glass and two layers of cement. All the vertical tanks are shorter than the inside depth of the frame by an amount representing the thickness of base glass and two layers of cement. If pieces of stiff cardboard are cut to the estimated dimensions the calculations can be checked by fitting the pieces within the frame. If the pieces are taken to the glass supplier he will cut the glass exactly of these patterns for you. To glaze the frame, first spread the cement

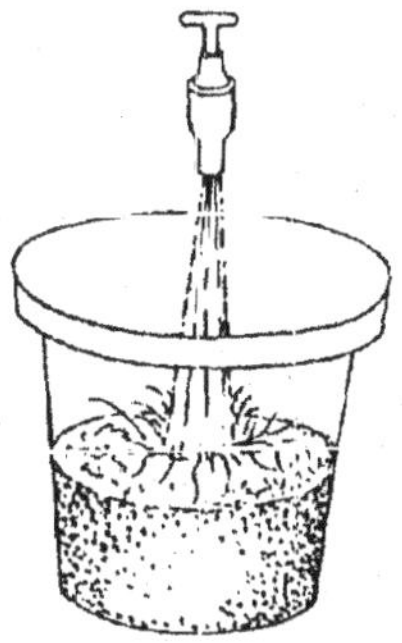

Fig. 1: Gravel Washing

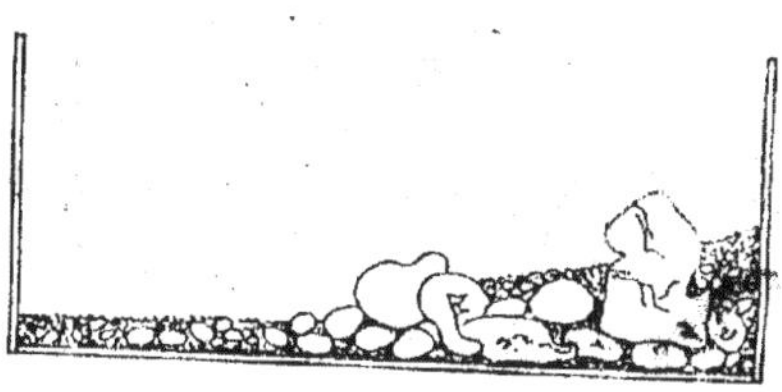

Fig. 2: Rocks Arrangement in Tank

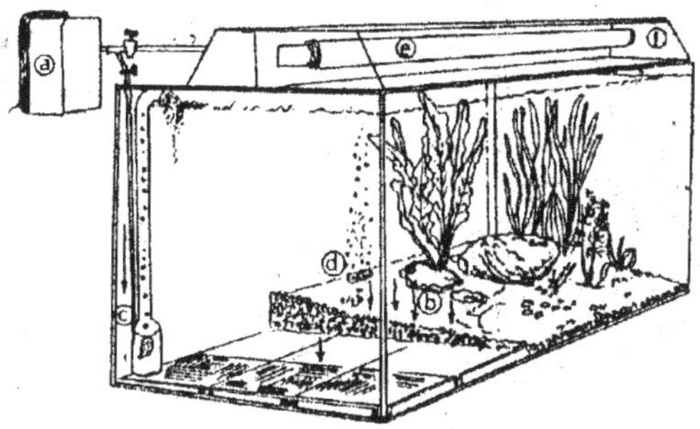

(a) Air Pump
(b) Draws Water Through Base Filter
(c) Air Lift Water Pump
(d) Diffuser Air Stone
(e) Light
(f) Reflector Hood

Fig. 3: A Closed Aquarium System

with a knife in the angles of the frame to form a layer at least ¼ inch thick. Each panel in turn, in the order given above, is thin pressed, into the angles with both hands exerting an even pressure so that excess of cement extrudes between glass and frame. Trim the cement extrusions with the knife. When all the glasses are in position and it has been checked that there is good contact all round, the tank can be filled with water. The weight of the water will serve to press the glass home and maintain it in place, and any possibility of harmful materials that dissolve out of the cement will be removed when the water is discarded after the aquarium has soaked dried for a day or so.

In all glass aquarium, the glass panels, cut to size, are cleaned and dried thoroughly and then they are cemented by their edges with silicone aquarium sealant, which is sold in tubes. Silicone Goldsize aquarium sealant is also useful if your aquarium ever starts to leak water. In such an eventuality the tank is emptied and after it has dried, sealant is applied around the inside angles where the glass panels meet. When aquarium is ready decorate it with rocks/ stones/pebbles (Fig. 2) and put plants and light (Fig. 3).

Filtration

Filters have several roles one is mechanical, involving the removal of inorganic and organic particles from the circulating water. Another is biological, the most important aspect of which is the conversion of toxic ammonia into much less toxic nitrate. A further function is the adjustment of pH.

Aeration

Vibrators, bubblers, aerators are available in market can be purchased. Introducing a stream of air bubbles into water ensures a local saturation with dissolved oxygen. Aeration is also useful for removing gases from super saturated water or for preventing super saturation, for instance when water is being treated.

Temperature Control

Thermostat with regulators can be used. The most common refrigerants are expanding gases. However, more stable temperature control can often be achieved by the use of a liquid secondary refrigerant.

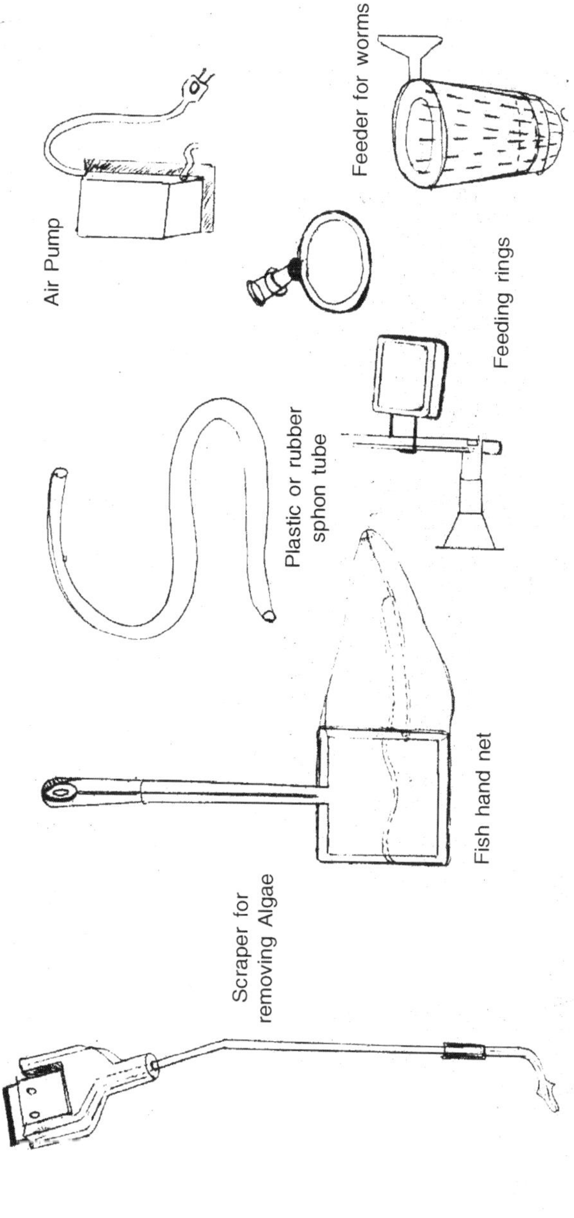

Fig. 4: Devices for an Aquarium

The design of a versatile temperature control system, capable of heating and cooling, and maintaining the temperature constant with in narrow limits is a matter for the specialist.

Pipes and Valves

The pipes most commonly used in aquaria now are unplasticized polyvinyl chloride and acrylonitrile butadine styrene. The commonest types of valves used in aquaria are ball diaphragm, gate, butterfly and needle type.

Conclusion

It is not only difficult to lift the filled aquarium or to change its position, but to do so, may cause a leak to start. Because of its weight it should have a firm and strong support. Angle iron stands are sold especially for aquaria, although a well joined wooden frame will serve just as well, and can be used to form a cupboard beneath the aquarium.

A position to one side of a window will be suitable, situations where the aquarium receives no light at all and those where it might be subjected to the full strength of the afternoon sun should be avoided.

The water flow through the filter should be kept constant since the microbial population will have a certain oxygen demand which must be met. Conducive environment is very necessary for aquatic inhabitants in aquarium.

References

Evans, Antmony. Aquaria.

Hawkins, A.D. Aquarium Systems.

Stephen, Spotte. Sea water aquarium.

Chapter 2

Construction, Decoration and Necessities of Aquarium

Before any fish is aquaired, some attention should be given to its accommodation. The globular fish bowl should not even be considered because it is invariably too much small and does not provide a sufficiently large water surface to fish, indeed, any glass aquarium. The first question that should be asked is whether to buy an aquarium at all. True, it is very easy to pop into nearest pet shop and buy an inexpensive aquarium off the shelf. However, there is an alternative i.e. you can build your own, often (but not invariably) for less that it would cost to buy one and the skills involved are well within scope of vast majority of potential aquarists.

Basically, all that is required is five panes of glass, a tube of silicone (rubber) sealant (most include instructions) and a certain amount of common-sense.

Here a table is given for sizes of panes of glass required for a variety of tanks.

Tank Dimensions (inches)			*Glass thickness (mm)*	*Water capacity (l)*
Length	*Width*	*Depth*		
18	10	10	4	27.3
24	12	12	6	54.6
36	12	15	10	91.0
48	12	15	12	196.5
60	18	18	13-14	318.5
72	18	18	16-18	364.0

This is quite an essential piece of information because the sizes of panes must allow for thickness of glass used. For example 24″ × 12″ × 12″ tank does not consist of 24″ × 12″, 24″ × 12″ and 12″ × 12″ sheets of glass only. The bottom, front and back are full size, i.e. 24″ × 12″ (bottom) and 24″ × 12″ (front & back) but the two ends need to be smaller than 12″ × 12″ to allow them fit comfortably and accurately into their allotted space. The actual overall size will of course, depend on the thickness of glass used.

All exposed edges need to be properly levelled, otherwise serious injury may result. Provision will also need to be made for a suitable aquarium cover or hood. This can be taken the form of a few simple "spots" glued on at strategic points along the top inside edges of tank (or on the under side of cover itself, if this consists of a glass sheet). Alternatively, an all round inside edge may be smoothed and stuck on with silicone sealant. This will allow a suitable sized cover hood to rest comfortably inside the lip of aquarium minimizing the risk of accidents.

Aquaria may be commercially produced, on the whole to be more presentable than home made ones. If the most expensive and luxurious models are excluded, then factory built aquaria compare favourable in price with home built ones and this may well make the time and effort spent in construction not too attractive proposition. In addition, these aquaria are produced in a range of standard sizes to meet the needs of most aquarists. Many models come with specially designed hood which lessen the burden on the aquarist even further. In this way one may open a small scale factory of aquarium with novelty of designs of moulds and supply it to aquarist.

There are usually three types of commercially produced aquaria.

Perspex Aquaria

These aquaria are sometimes constructed with bowed front while giving the tank an attractive appearance can cause a certain amount of magnification and distortion. Perspex glass also has other disadvantages in that it can yellow with age and easily scratched. However, Perspex is cheaper than glass. Bearing in mind the other inevitable expenses involved in starting, this must not be overlooked. With care, Perspex aquaria can look good and last for many years.

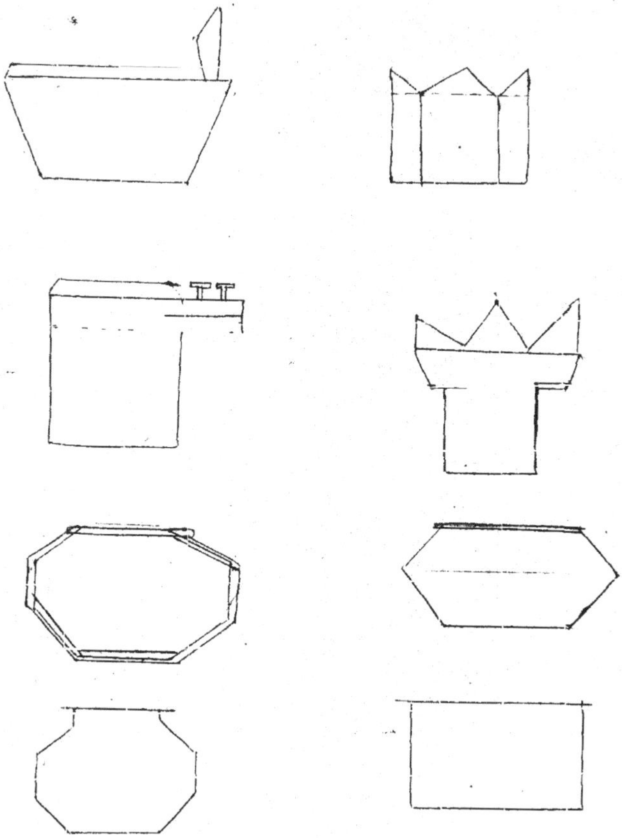

Fig. 1: Different Designs of Aquarium

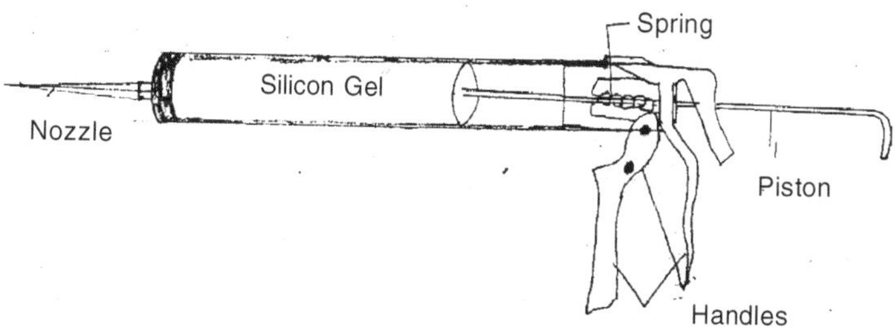

Fig. 2: Aquarium Sealing Gun

All Glass Aquaria

Sealing with silicone rubber with the help of gun gives good finish. The gun pouring the silicon rubber over the glass edges is handled carefully and then pressure is applied for setting of glass to avoid leakage. These are, by far, the most popular aquaria in the market (liked by customers). The quality of silicone sealant used now-a-days in such that all glass aquaria measuring 7 feet in length or more are constructed at economic prices. To prevent bowing, large aquaria of this type have one or more strategically placed strengthening struts between front and back. Most silicone sealants last, at least, ten years, or in many cases even longer (Don't use M-seal locally available sealant, it is poisonous for fishes) silicone sealant is costly a little bit. One tube is sufficient for 5-6 aquaria, when one tube is opened, make the 5-6 aquaria immediately otherwise it will become useless.

Framed Aquaria

In past, most aquaria of this type consisted of five sheets of glass stuck on to an external frame of angle iron. Whilst being extremely strong and providing excellent protection knocks, the angle iron frames soon rusted, despite the use of anti rust paint with unsightly and potentially dangerous consequences for the tank inhabitants. Today there is a range of superior coatings plus the use of substitute materials in place of angle iron which have played an important role in restoring formness in framed aquaria.

It must be stressed that a well constructed and well protected framed aquarium such as some of the models which are currently improved should be sold in the market. It is as good as any other type and safer and more attractive.

Wooden frames can also be made from hard wood such as oak and these frames are glared with water proof of steel, alloy etc. materials just like metal frames. They form highly decorative aquaria and are perfectly satisfactory if the joints are strongly made.

Before painting the frame should be thoroughly cleaned with a wire brush and made smooth and bright with filer and emery paper. The inside top frame of the tank is liable to damage or cause damage so this part should be given special attention.

Now place the glass panes. The base glass is always fitted first. In place of glass base, asbestos sheet of same bottom size over a tin foil may also be used. This will provide a strong base. Its dimensions need to be only slightly less than those of the inside of the frames. Panes at far ends will be fitted in last between the front and back panes (after smoothing their edge with linseed oil) so their width need to be reduced by an amount equal to two thickness of the glass and two layers of sealing compound.

Sealing compounds are:
- Silcone rubber
- Bitumen (No. R/85/25)
- Putty

The use of silicone rubber is similar as in case of all glass aquarium. Bitumen the battery sealing compound is first melted then it should be poured in a narrow mud canals made at the corners where sealing is to be done. This is for the perfect and localised even pouring of bitumen. It is cheaper. A little carelessness can cause great damage, precautions should be taken. They are as follows :

Precautions

(a) Melting should be proper

(b) Handle, with care otherwise glass will crack or you will be burned. Give full attention to your work. Any diversion may cause loss to you.

(c) There should not be light patches of bitumen at sealing ends.

Disadvantages

(a) The age of aquarium is not more.

(b) It is a hard sealing compound and creates difficulties at the time of leak repair.

  (i) It is tedious to scrap the hard set bitumen.
  (ii) Readymade melted bitumen is not available.

Poor adhesion may be indicated by lighter patches where air has trapped between any sealant putty and glass. Glazing compound should have four qualities:

Water resistance
- Adhesiveness
- Elasticity
- Being non-toxic/Non toxicity

The elasticity is required by glazing compound to allow the slight movement of glass due to water pressure and fluctuations due to temperature. The hard setting cements should not be used because, they don't take up slight movements and they are difficult to remove if it is ever necessary to repair.

The best quality of linseed putty is easiest to use, quantities should be well kneaded with hands quite soft, then pressed liberally into the frame with fingers, ensuring that corners are well filled. A cement made of equal parts of zinc oxide and red lead in double boiled linseed oil. Another type of cement is made by mixing two parts red lead, two parts white lead and one part putty using minimum amount of 'Gold size' (adhesive compound) to write the mass. The lead and putty can be mixed easily by laying them on a board, working them together with a broad bladed putty knife. A simple cement can be made from equal parts of white lead ground in oil and whiting. The lead and whiting are mixed together and small quantity of boiled linseed oil is added until cement attains a dough like consistency. The dough should then be hunged in a muslin bag and left overnight to settle and then given a final kneading before its use. This particular formula has the advantage of remaining soft and in good conditions for quite a long time if stored in a airtight lid. Consequently it is a useful standby for emergency repair. A quick drying cement formula: 16 parts whiting + 4 parts litharge + 2 parts powdered resin + 2 parts powdered red lead may be used. In this the ingredients should be first mixed, then a good quantity of varnish may be added until the mass attains consistency of still dough because it harden fast, it is advisable to make it only in small quantities and use as required. Plaster of Paris should never be used as an ingredient in any cement as it has a tendency to make such cement dry and brittle.

Depending on the cement formula used, it may be necessary to age the tank in a warm room for a few days or a few weeks before using it. The cement remains semi soft and workable for some times so that it will give a little under the water pressure and ensure a water tight seal. When filling a new tank of 20 gallons or greater capacity, it is advisable to fill it half-full first; wait twenty-four hours, and then fill it one-quarter more; then fill it completely after another twenty-four hours have passed. Minor leaks in a new tank usually seal themselves in a short time. Excess cement should be squeezed out between the glass and frame. Do not under any circumstances force the cement back under. The trimmings are good that cement can be used rarely.

After making a tank, always test the cement with your hand or finger nail. The cement should be fairly soft, if it will not yield to your finger nail. If the tank is old and it may leak when filled. It is not usually necessary to cement a tank in order to repair a leak. If the leak is a minor one, try to stop it by stirring with a handful of mud into the water. Often the particles will lodge in the crack and stop the leak. If this method fail, it will be necessary to empty the tank and allow it to dry thoroughly. Clean all the gravel, dirt and excess cement from the inside seams, both horizontal and vertical; with a screwdriver or strong knife. Liquid aquarium cement, available at most dealers can be used to seal the leak. Tilt the tank on one corner. Pour a little of the liquid cement in at the top so that it runs all down the seam. As it collects at bottom, tilt the tank so that the cement runs as long the bottom seam. If necessary, add a little more cement. Repeat this at the next corner and so on until all the inside seam have received a liberal coating of cement. Blow this to dry for several hours and repeat the process. It is better to cement all the the inside seam rather than just the spot where the leak appeared.

Frequently a leak will start inside and run along between the frame and glass to appear at an entirely different place outside. After drying for twenty four hours the tank can be filled. If you have done your work properly, the tank will be watertight. A minor crack in the slate can often be stopped with the same liquid cement. Paint is liberally done in and around the crack. Then set a piece of glass over the cement covering the crack while the cement is still wet. The plastic cement is not as satisfactory for use in repairing leaks. Glacial Plaster of Paris or tar are not used in sealing aquaria.

The replacement of a broken aquaria glass requires a different procedure. For this turn the aquaria upside down and run a hot electric iron over the metal frame. Do not hold it in one place but keep moving it. The object is to heat the cement enough to soften it without cracking the slate or glass. Experts use an acetylene torch, playing it back and forth. Keep pressing the slate firmly until it loosens and can be removed. If it is an end glass which is broken, pick out the pieces and clean the frame carefully, properly replacing it. When one of the larger glass is broken, it is necessary to remove the two end glasses first, using the same technique as for the bottom. To replace the glass, work plastic aquarium cement between your fingers until it is soft. Adding a few drops of linseed oil, or heating, helps soften the cement. Roll it between your palms like a rope and press it in to the frame all around. Place the upper edge of the glass under the cover hanging upper lip and press the bottom into place. Fit in the two side glasses first, then the ends and then bottom. Heavy wired glass may be substituted for a broken slate.

It can produce a satisfactory tank more cheaply than one can without such facilities. However, do it yourself, you might find it interesting to buy just the frame (dealers have them), then cut and set the glass himself.

Here are the sizes of aquaria in gallons and inches that are usually available in stainless steel :

Capacity (Gallons)	*Length (inches)*	*Width (inches)*	*Height (inches)*
2	10	6	8
3	12	7	9
5	14	8	10
5½	16	8	10
8	18	9	12
10	20	10	12
12	20	10	14
14	20	10	16
10	24	8	12
12	24	8	14
14	24	8	16
15	24	12	12
17	24	12	14
20	24	12	16
20	30	12	12
23	30	12	14
26	30	12	16
29	30	12	18
30	36	12	18
34	36	12	18
39	36	14	18
42	36	16	18
50	36	18	18

Size larger than these must be made to order, usually with an angle iron frame. Odd in between sizes must also be made up specially. Usually the charge for a special size is three or four dollars (Rs.200-250) more than the next larger size.

Marble tanks are usually made only in 2½, 3½, 5½, 7½, 10, 15 and 30 gallon sizes with the same dimensions as the equivalent capacity stainless steel frame tanks.

Frequently tanks with the same surface dimensions are manufactured in varying heights. In very few cases have the proportionate height been carried to dangerous extremes. But, other dimensions being equal, the lower tanks can support as many fish as the, higher one. Increased heights have an aesthetic appeal for many people.

Advise for beginners way or to convince the customers

Choosing the proper tank size

It is very common and proper for the beginner to approach or cautious about his first purchase of an aquarium or manufacture of aquaria. The usual advise is to buy a small aquarium and see how it works out. If the experience is satisfactory, he feels, he can then invert in a larger one. On the surface, this appears to be the wisest course, but close examination of several certain things are as follow. The equipment and accessories necessary for the tank, pump, filter, heater, thermometer and other essential accessories for the tank and room are designed for an average size aquarium and are priced accordingly. It costs almost as much for them when setting up a 2 gallons tank as for a 10-15 gallons one. Even more important is the difficulty of maintaining a small .aquarium. Everything else being equal a 2½ gallon tank is harder to maintain properly than a 15-gallon tank. The small amount of water becomes polluted too readily. The temperature fluctuates too much and it becomes over-recorded almost before the hobbyist has even started collecting. A small tank is proportionately more expensive and really a job for an expert, not an amateur. A beginner who starts with a small tank all too often returns to the dealer with in few weeks to find what kind of a trade in; he can get on his little tank for a larger one. As his tank is second-hand by them he must, of course, take a loss. We recall one beginner who resisted all advise and insisted on starting with a 2 gallon marble tank. He finally wound up after 5 trade-in with a 35 gallon stainless framed steel tank. Counting financial losses on the trade and not taking into account time and trouble, he paid exactly twice as much as he would have if he had, purchased the larger one first.

One advise to the beginner is to start with a tank between 10 and 20 gallons, the 15 and 17 gallons sizes being the best. These sizes are not too expensive, they are small enough to handle fairly easily and yet large enough to house a good looking and fairly representative collection. Then too if the hobbyist, should decide to go on to bigger and better aquaria, the medium sized tanks are always handy and have a space breeding, hospital, nursery or isolation tank.

Larger tanks, more than the 20 gallon size pose problems in balance which are difficult for the beginner to cope with. In addition if anything go wrong, it is quite a chance to change and clean the larger tank. An aquarium made to order for a special location always bears the depth of surface, keep it mind.

For offices, always suggest, larger aquarium, larger the aquarium, the smaller the risk of causing serious imbalance. This is because larger volumes of water, particularly when combined with healthy plants growth and adequate filtration can absorb, errors of judgment much more easily. In other words, larger the volume of water, greater is its buffering capacity.

Setting of Sand and Gravel

For nice setting first cover the bottom of aquarium with a layer of clean sand white or coloured in two inches height you can design the sand bed setting according to your/customer aesthetic sense or as showed in diagrams. The sand can be obtained from market or from beaches. The sand supplied as ready wash for aquarium by shops is convenient to use. But any sand which is having coarse particles will be suitable after first watch. About 12-18 pounds of sand will be right quantity for 24″ × 12″ × 13-15″ size of aquarium. Wet the sand first and stir it up in a bucket with water several times before placing it in aquarium it is better to wash with 0.1% $KMNO_4$ or lime and put it under flowing tap water but for white sand give a wash of 0.1% salt solution and wash under continuous flow of water several times. It is better to form a layer of sand first and give some height at the back and sides and also give a slope at front. It will encourage any bits and pieces of unwanted material to roll towards the front glass and from this position it can most easily be removed. The medium gravel with grains of small size 'O' in English printing is considered as best. If it is too fine plant roots will have trouble in pushing their way through it. If coarse gravel is used fish detritus and uneaten food will settle down through it, resulting into foulness. Take enough of these gravel to cover one inch bottom with a gentle pouring after using disinfectant and thorough washing with fresh flowing water. Push the nozzle or stream of water around the gravel and loosen the dust fine particles to overflow water. When water comes clear and transparent your washing is perfect. Rearrange the gravel with your desirable design. After bed (sand and gravel) you can use white paper over the gravel and pour the water gently or a cup in the bottom letting it overflow into the paper so not to disturb the gravel setting. The quality of sand/gravel causes the efficiency of aquarium. The fine sand is apt to go black owing to its habit of packing down hard and so giving bacteria a chance to develop with larger green sand there is a certain water movements which keeps it clean. In any case if the sand go black it will be restored to its original state if it is placed in sun to dry. This disruptive procedure will necessary if coarse sand is used and decayed matter particularly rottening of plants is removed at once. Naturally black sand is used by some of the continental aquarist but it depends upon the colour of the fish as well as the choice of customer.

One can use rocks/drift wood, logs bamboo sticks, shells dead corals, toys any thing which you feel can enhance the beauty of your aquarium without changing the water quality (by leaching) or any harmful effect.

Light

Infra red lighting in combination with an infra red viewer, is effective only in very shallow water. The blue end of the spectrum offers even less scope for viewing fish. Blue lies nearer the peak sensitivity of most fish and the short wavelengths are readily scattered leading to poor resolution. An aquarium room light is useless. Electric light should be immediately over the tank.

Incandescent bulbs give out more red light which makes for the vigorous plant growth. Fluorescent light do not cast dark shadows but the ordinary day light tube it too blue to have much effect on the plant life. We should use warm white tubes placed on the front side of the aquarium as this display to the fishes to their best effect. There is no hard and fast rule as how much wattage of light should be provided. By trial and error, one can provide a light according to the condition of the room and the climatic condition of the area. Many species prefers dim lighting. Some fishes prefer regular light and some interruption light. For many species maturation is triggered by changes in the day length. For 18″ inches length wattage of filament bulb 40 watt, wattage of fluorescent tube 8 watt.

For 24 inches length	2 × 40 watt bulb 15 watt tube
For 30 inches length	2 × 60 watt bulb 20 watt tube
For 36 inches length	3 × 40 watt bulb (3 bulbs) 2 × 20 watt tube
For 48 inches length	2 × 30 watt tube 3 × 60 watt bulb

Larger tanks those lifted in the table can be lit by dichromic spot lights mounted over the tanks. The only cover required in the top. Ordinary tungsten filaments are cheaper to buy than flourescent tube.

Heavy starter units are avoided but the usual side way mounting of these bulbs in a standard cover give a short light. Lighting the case while the lamps are lit on often breaks the filament. The bulbs also heat the surface of water to unacceptably high levels for many fish. Such as cold water species or family Anabantidae (*Anabus*). The heat can also create condemnation problems but this and the surface heating can be overcome by placing a glass sheet between the lamps and water.

Fluorescent tube have the advantage of being cool and cheaper to run. The started unit is the best housed on a wall behind the tank, rather than on or in the cover. This is because its heavy weight creates a danger of the whole lids. There must be stainless steel or nylon coated steel, electric connections can be smeared with silicon sealer. Domestic wiring is sufficient for all normal aquaria and hence a wide range of readily available accessories can be used. Each tank have its own lighting arrangements and they may lit from a single circuit with a master switch that can be controlled by a timer. The lighting of the individual tank can, of course, be controlled by separate timer, if required.

Air Pumping Systems

There is probably no field where the aquarist is offered so wide a choice of equipment or is promised so much by the manufacturers. Each system has its particular uses and applications. Your requirements will vary according to the number and sizes of tanks in operation, the type of fishes or invertebrates you intend to keep and the degree of sophistication you are trying to achieve. This review of the available equipment should help you decide which system suits your needs and pocket. Reliability of the system is of foremost importance an inadequate air supply at a critical time can be heatbreaking. The air pumps currently available fall into four principal categories: (1) diaphragm pumps, of innumerable models and outputs; (2) piston pumps; (3) carbon blade rotary pumps; (4) low pressure air blowers.

Diaphragm Pumps

This first category is the choice of probably the vast majority of enthusiasts and the reasons can be summarized quite easily. The performance and reliability of these pumps is good, they are neat and small, and are reasonable in price. All diaphragm pumps work on the principle of a small vibrating diaphragm valve which is caused to open and close by passing alternating current electricity through a magnetic coil. Each closing of the valve causes a bellows effect and the rapidity with which this occurs produces a continuous stream of air through the outlet. The simplicity of the design results in a degree of reliability unusual in a mechanical product and they can be left to run unattended for months and even years without any maintenance. The only part that wears out eventually is the diaphragm, but spares are inexpensive and readily available for reputable makes of pump. Diaphragm pumps do vary, however, in their capacity to produce air, and also in their price and amount of noise they produce. As most aquariums are installed in a living room, this noise level is probably the over-riding factor to be considered when making a choice of pumps, a point which if ignored may cause more domestic conflict than any other! Always insist on listening to the pump operating under normal conditions in the dealer's showroom before making your choice, few shops will be willing to exchange a pump simply on the grounds that you have subsequently found it to be too noisy. Some aquarists find even the most silent diaphragm pump too noisy. It is often helpful to stand

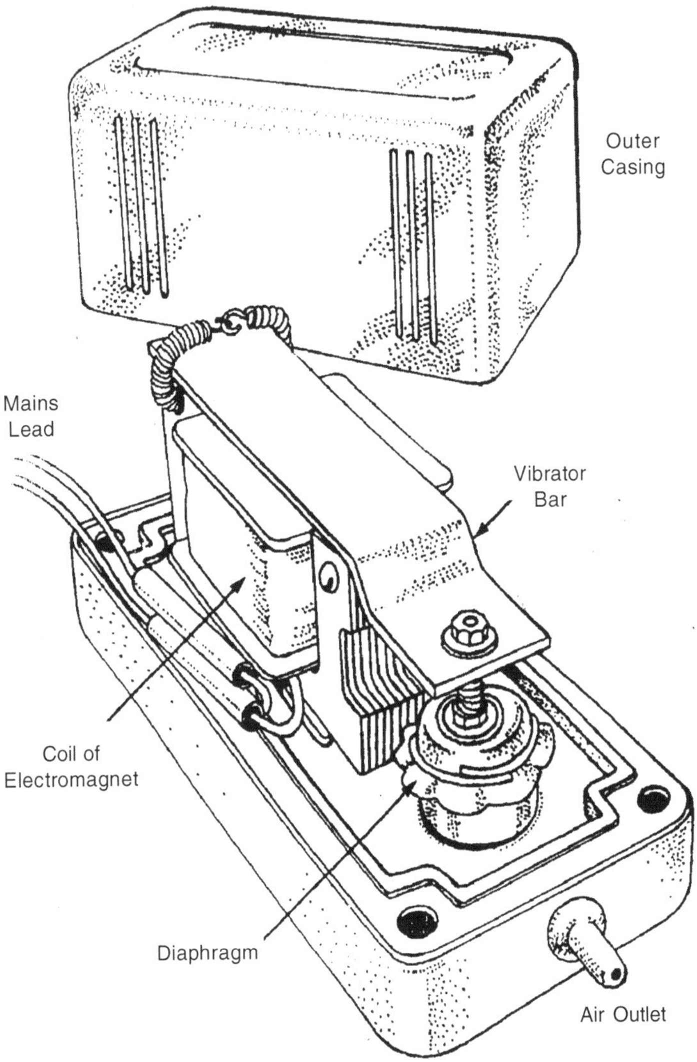

Typical Diaphragm Pump

the pump on a small piece of sponge rubber which will help to deaden the vibrations. If this can be combined with placing the pump in a cupboard or box, it should be possible to reduce the noise to below normal audible levels. Many of the models offered today have more than one air outlet and this increased capacity is essential where a larger than average tank has been selected. Do bear in mind, however, that there are several high capacity models available which have only a single outlet and the simplest way to test the capacity of a pump's ability to deliver air is to place an air tube from the nozzle into increasing depths of water with an air diffuser attached. Weaker the model, the less able it is to displace the water and as the depth of operation increases so will the efficiency of the weaker pumps decrease. Do not select a pump which cannot push air through a column of water twice the height of your selected tank. Should you contemplate the possibility of later expanding to more than one aquarium, it is false economy to buy a pump which will serve a single tank. A really good pump can be used for several aquariums and will usually cost less than several smaller pumps.

To sum up, the intending purchaser should select a model which his dealer both recommends and is prepared to demonstrate; he should choose a capacity which is greater than that required rather than less, make certain that spare parts are stocked by his dealer, and compare the pump in price, design and noise output with other models. Finally, remember always to place a diaphragm pump at the same level as the aquarium or above it because any interruption

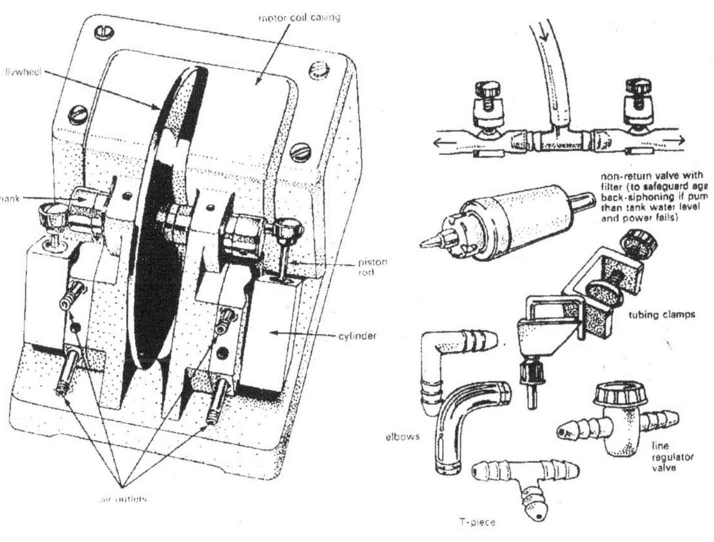

Piston pump

in the power supply may cause a siphoning effect to take place, with disastrous effects to the pump.

Piston Pumps

This type of pump seems to be made only by two manufacturers, one in the USA and the other in Britain. The advantages of piston pumps are that they are extremely quiet, very reliable and hardly ever wear out. Their drawbacks are that they are certainly not as cheap as diaphragm pumps and need an occasional application of oil to the pistons to maintain them in working condition. This lubricating oil must be prevented from entering the aquarium through the air line, because even the smallest amount of oil on the surface could cause major problems. A cotton wool plug filter placed in the airline will mop up minor amounts of oil or other dirt in the air. The major problem, from the point of view of the hobbyist, is the limited supply of piston pumps. You may have to pester your dealer to obtain one from his distributor. They also cost more but because of their reliability and low noise are highly recommended.

Filters, plants, fishes and their feed are also important discussed in later chapters.

Chapter 3

Marine Aquaria

Marine Tanks

Unless you are using natural sea water, an artificial sea salt must be added to the aquarium water after you have brought it to the right temperature. Read the instructions on the packet and add the salt gradually, checking the concentration of salt water by measuring its specific gravity with a hydrometer. Hydrometers are sold by any good dealer and are an essential piece of equipment for the marine aquarist. When the salt has fully dissolved and the temperature had reached 75°C F (24°C) the hydrometer should read exactly 1.022. You must now mature the filter bed of the aquarium and this best done by applying a dose of freeze-dried bacteria to the water and then feeding the bacteria with infusoria. In this way it is possible to mature a marine tank in eight to ten days; any other method will not only take several weeks but will allow the multiplication of unwanted competing bacteria which although not intrinsically harmful to the fishes or invertebrates will reduce the quantity of livestock the tank can maintain. The test the maturity of the tank you should use a standard nitrite testing kit which should be capable of measuring down to 0.25 parts per million (ppm) on its colour scale at which level it is safe to introduce the first fishes. At least two kits on the market do satisfy this condition but some do not. Any level above 0.25 ppm is dangerous for some sensitive species. You will find that the upward curve of the nitrite cycle will reach its peak at about 8 ppm and will then fall to 0.25 or less when tank is mature. Do not rush to introduce several fishes at once and water the nitrite level daily. The nitrite level often goes up a point or two after the first fish is placed in the tank so make sure that this fish is one of the less sensitive species as a damsel. The nitrite concentration should drop after 24 to 48 hours and once this has occurred you may introduce

The assembly and connection of the basic equipment

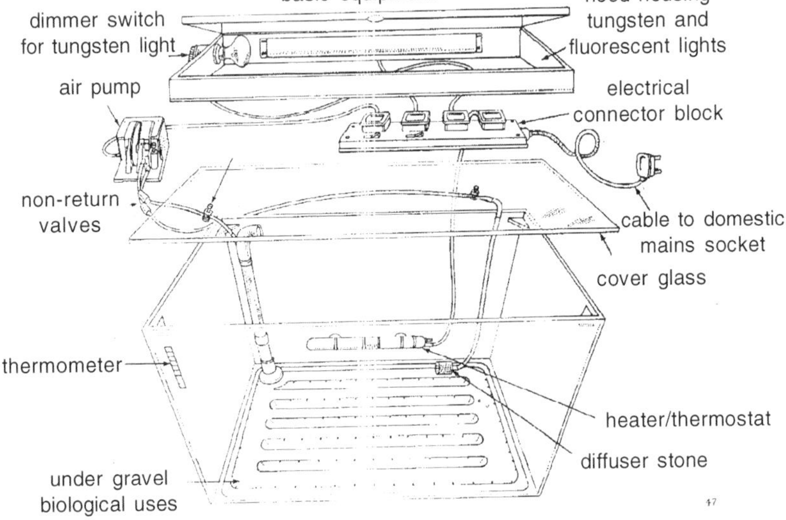

further fishes, one or two at a time, but as precaution continue to monitor your nitrite level for the first month. A well matured system should register a zero reading at all times after the first few fishes have been introduced.

The Tank

The traditional aquarium that has existed for many years has rectangular iron frame with glass sides embedded in putty. High quality stainless steel frames are excellent for all types of aquaria. However, if putty is used, the inside corners must be silicone-sealed for seawater usage.

Some modem tanks have metal frames sealed with a nylon or polyurethane coating prior to glazing. These tanks are suitable for all uses, but again if putty is used, the inside comers must be silicone sealed for seawater.

All-plastic tanks have been popular in recent years because they are non-toxic, light and require no maintenance. The rising cost of plastics, however, has made these tanks expensive and some manufacturers have changed back to glass. Small plastic tanks (2-60 litres) manufactured from a single mould, are useful for handling fry and for quarantine and treatment purposes. A problem with plastic tanks is that the surface can be easily scratched making the sides opaque, so abrasive cleaning materials must be avoided.

The ideal tank is probably the all-glass silicone-sealed unit. This is simply a glass box and can range in size from 10 to 2000 litres. Small tanks of up to 100 litres can be made from glass of 6 mm thickness, but 8 mm glass is preferable. The front, back and base can be 6 mm thick with all the side 6 mm. For tanks of more than 100 litres capacity, 8 mm glass should be used throughout; while for the tanks of more than 200 litres, 12 mm glass is necessary. Large aquaria of 1000 litres need 24 mm glass which must usually be specifically ordered. There are variations, you are making aquaria for marine inhabitants.

Silicone sealer can be obtained in small tubes or cartridges, the latter requiring a specially designed chalking gun. Glass should be bought in a cut and polished state and the edges cleaned with acetone or a degreasing agent prior to sealing. The prime need of marine aquaria is sea water. Synthetic sea water powder is available in the market.

Heaters, Thermostats and Thermometers

A combined heater/thermostat should be sited according to the manufacturer's instructions, usually vertically in a near comer. If an aerator is placed near the instrument, the water flow helps to dissipate the heat. A separate heater may be placed horizontally on the gravel; this position aids convection. Extra long heaters are available which have a low heat content per unit area. This allows the heater to be buried beneath the gravel without the danger of the glass cracking. The advantage is only an aesthetic one, however. For a similar reason, there are under gravel flexible heaters and under tank heating blocks available, but the standard heating element in a glass tube is recommended for ease of use and maintenance.

This is particularly important with marine tanks where salt crusts develop with age; these creep away from the tank on any surface and work into the electrical connections giving rapid corrosion and electrical short-circuits. For the lighting circuit of a marine tank, the author runs a cable through polythene tubing sealed with silicone sealer at the connection to a fluorescent tube and at a point at least a metre away from the tank.

Each tank needs a thermometer to monitor the water temperature. The temperature of the water involved in water changes or fish transfers should always be checked with a thermometer because rapid temperature changes will severely stress the fish.

Aeration

Continuous aeration is very good husbandry since it mixes the water, supplies oxygen for the fish, and removes carbon dioxide. Many cheap air pumps are available in the hobby trade, though they are often noisy, are of limited power and many frequently fail. If only one or two tanks are required, such vibratory diaphragm pumps are acceptable, but a spare pump and several replacement diaphragms should be stocked. The pump should be mounted above the tank level or the air-line fitted with a non-return valve to prevent back-siphoning when the pump is stopped or fails. A loop in the air-line 8 cm (3 in) vertically above the tank's water level will also prevent back-siphoning by absorbing the oscillations when the air-flow stops.

Properties of Sea Water

One can purchase and prepare sea water or set a pipeline with graded filters directly from sea to aquaria or by dissolving different salts one by one in water according to sea constituents.

Sea water has many characteristics that are similar to pure water, but there are also certain differences. Sea water is 96.5% water and 3.5% salts. Of the salts dissolved in the sea water, sodium (Nat) and chlroine (Cl) make up about 86% of the ions and sulphates (SO_4^{-2}), magnesium (Mg^{+2}), calcium (Ca^{+2}) and potassium (K^+) compose about 13%. The rest of the ions identified in sea water combine to make only 1%.

Temperature

Temperature control is an critical aspect of good aquarium management, as such slight temperature change may have harmful effects on the animals being maintained.

(a) Cooling

In seawater aquarium applications, the evaporator coil, which are copper, must be coated to protect the animals and plants from copper poisoning. The evaporator coils can be placed in the water system anywhere adequate circulation is available to augment heat exchange. One method is to place the coils underneath the filter plate of the bacteriological filter so that all water is cooled as it leaves the filter and before it is returned to the surface of the aquarium by airlift pumping.

(b) Heating

Nowak (1974) described a solid state temperature controller for both heating and cooling. The project here is limited to heating seawater with immersion heaters. Immersion heater must be grounded properly or they are potential safety hazards, particularly the larger units that require more than 120 volts for normal operation. In addition, no immersion heater should be operated without a thermostat to control temperature.

Water and Salt Balance

Aquatic organisms are solutions within a solution. One of the major task is to balance physiological water and salt while simultaneously taking in seawater. Most are stenohaline and can tolerate only limited changes in the ionic strength of their environment. Fluctuation in the external salt concentration disturb normal water and salt regulation in the tissues, resulting in physiological stresses. The mechanism controlling water and salt concentration are diffusion and active transport. Osmo regulation in the overall process by which water and salt balance is achieved.

Gas Exchange

Gas exchange in a sea water aquarium takes place at two locations – (1) the air water interface; and (2) through air bubbles rising in the eduction pipes of airlift pumps.

(a) Air-Water Interface

Gas exchange between air and water is a function of the differences in partial pressure of a gas in the atmosphere versus the water, and thickness of a surface film located at the air-water interface. This film, formed by surface tension, is known to oceanographers as the laminar layer.

The movement or flux of a gas through the laminar layer depends on differences in partial pressure between air and water, and also on the "diffusivity", or rate of diffusion of the gas as shown by

$$\text{Flux} = \frac{\text{Concentration} \times \text{diffusivity}}{\text{thickness}}$$

where, concentration = partial pressure × solubility oxygen and nitrogen, the most common gases in the atmosphere, have the similar diffusivities i.e. about 2×10^{-5} cm sec^{-1} at 20°C.

(b) Rising and Bubbles

Air bubbles driven beneath the surface of the sea by wave action are important in oceanic gas exchange. Similarly bubbles rising in the eduction pipes of airlift pumps greatly augment gas exchange in sea water aquarium. Two of the controlling factors are the same as in gas exchange at the air-water interface: differences in partial pressure and the presence of laminar layers. In air bubbles released beneath the surface of the water, hydrostatic pressure becomes still a third factor.

Kanwisher (1963) reported that small bubbles of 0.05 cm in diameter were rapid gas exchangers. Bubbles of this size and with laminar layers 0.0010 to 0.0015 cm thick, were more than 70% equilibrated with CO_2 after rising through only 5 cm of water in 2.5 second.

Aeration Methods

Kils (1976/1977) tested three methods of aerating seawater, as determined by the dissolved oxygen concentration upto the saturation point. The equipment used consisted of (1) a simple aerator, in which air bubbles were released at the bottom (2) a venthri, in which air was entrained into a stream of water discharged beneath the surface and (3) a rotation aerator that utilized air bubbles pulled into water by turbulence.

The conclusion was that sea water required less input of energy to aerate than fresh water, and that warm sea water was easier to aerate than cold sea water of the same salinity.

Airlift Pumps

Gas exchange in aquarium is more efficient, if the surface is disrupted exposing a greater percentage of water to the atmosphere in a given period of time. This process is made even more efficient if air bubbles in a continuous stream are allowed to rise in the water column. Gases in the bubbles diffuse down a concentration gradient into the surrounding water.

Water in an aquarium can be circulated by mechanical pumps or airlift pumps, but only air lifts provide both surface agitation and an endless supply of rising bubbles.

Nutrient Balance (of elements)

The buffer system of sea water is composed mainly of inorganic carbon compounds. In aquarium seawater, though carbon originates from the diffusion of CO_2 into the water from air bubbles or across the air-water interface and from the respiration of animals, algae and bacteria.

Phosphorus

Phosphorus exists in seawater in three states: dissolved inorganic phorphorus (DIP, also called as phosphate, orthophosphate and reactive PO_4^{-3}), dissolved organic phosphorus (DOP) and particulate organic phosphorus (POP). The last form is associated with floating detritus particles and less of responded microorganism. Orthophosphate is the most important form in water and the same is probably true in seawater aquarium. In sea water of p488, 1% of the orthophosphate is present as H_2PO^{4-}, 87% as HPO_4^{2-}, and 12% as PO_4^{3-} (kesterard pytrowicz, 1967).

Filtration

Marine fish require their water to be biologically filtered an on a small scale, this filtration is normally done by means of under gravel filters. A better system is to use an external power filter. The power filter is more efficient than under gravel one, gives a better water flow within the tank, and can easily be cleaned without disturbing the fish. Any filter medium with a large surface area is suitable, including silica sand, coral sand, gravel, glass beads and various resins. The author favours carbon and polyester wool. The carbon is not included for its adsorptive properties (which are quickly saturated), but for its large surface area. Since biological filtration relies on surface-active nitrifying bacteria the filter medium with the highest surface area will be the most effective. The carbon granules should be packed loosely in a mesh nylon bag, and to prevent them becoming clogged with solids, the bag should be sandwiched between polyester wool which removes gross material and gives a final mechanical filtration before the water flows back into the tank. Table is a guide to pump sizes that have been shown in practice to keep saltwater nitrite levels near zero.

Table 1: Power Filter Size for Saltwater Tanks

Tank Size	*Flow Rates/hour (h)*	*Examples of Filters*
Up to 501	Over 100 l/h	Eheim No. 2016 Martin No. 300
Up to 2001	Over 150 l/h	Eheim No. 2018 Martin No. 400 Nuova Turbo Filter
Up to 5001	Over 300 l/h	Eheim No. 2026 Martin no. 300
About 10001	1000 l/h	Eheim No. 2030 or No. 2034 Martin no. 300

Decoration

Most fish seem to prefer a tank decorated to resemble their natural environment. Any rocks, pebbles or ornaments placed in the tank must be non-toxic and sterilized with boiling water. Rocks must be lime-free to avoid increasing the water hardness, and should not have sharp edges. Granite, river- washed boulders or rounded lumps of glass are ideal. Many coral-fish need pieces of coral to weave through (these are usually bleached white when sterilized).

To give a better view of the fish, the back glass may be made opaque. An effective method is to cover the outside rear glass with a liberal coating of a white emulsion paint.

Lighting

Most fish prefer dim lighting conditions. The bright lighting of the aquarist's display tank is not necessary unless plants are present. Many fish are diurnal in habit, requiring a regular light/dark cycle. Wattage for an 8 hour (h) period of lighting where plants are present.

Table 2: Light Intensity for Planted Aquaria

Length of Tank	*Wattage of Filament Bulb*	*Wattage of Fluorescent Tube*
Upto 50 cm	40	8
60 cm	2 × 40	15
75 cm	2 × 60	20
90 cm	3 × 40	2 × 20
120 cm	3 × 60	2 × 30

Fluorescent tubes have the advantage of being cool and are cheaper to run.

Stocking Levels

The limiting factor to stocking level in the small aquarium tank is the oxygen content of the water.

To calculate maximum stocking levels in marine water aquaria, measure the water surface area in square centimetres. Measure the fish's length excluding the tail and group into small (up to 3 cm), medium (3 to 5 cm) and large (5 to 10 cm). Divide the surface area by 140 to give the maximum number of small fish by 240 to give the maximum number of medium fish and by 280 to give the maximum number of large fish.

Table 3: Stocking Levels

Surface Area (cm^2)	*Tropical Fish*	*5-8 cm Marine Fish*
700	10 (small)	3
1800	15 (medium)	5
2800	24 (medium)	7
4600	40 (medium)	12
8400	70 (medium)	20

Plants

In marine aquaria, few higher plants can be cultivated. The disadvantage of real plants is that unless conditions are ideal, they rot and pollute the tank. When conditions are good, growth can be vigorous necessitating constant attention e.g. pruning and thinning. But it is always suggested to use natural plants e.g. ulva, porphyra, Gracilaria, sea weeds, etc.

Foods and Feeding

The best basic diet for the small aquarium fish is a commercial flake food. The advantage of flake is that it will at first float for top feeders, sink slowly for middle feeders and lay discretely on the bottom for bottom feeders. Carnivorous fish have a higher requirement for protein and vitamins per body weight than omnivorous mammals and the usual mammalian foods (e.g. kitchen scraps) are nutritionally deficient. The ideal diet would be a live one such as

Artemia, Daphnia, Tubifex, Marine plankters etc. but such a diet is sometimes difficult to provide and can lead to the introduction of diseases or parasites. Sterile prepared foods are preferable. Most flake foods are made from the dried meals of fish, meat, cereals, etc. Because processing to a meal tends to heat damage amino acids, the total protein content of these foods is high to bring the limiting essential amino acids up to the levels required by the fish. A better balanced diet is achieved with lower levels of a higher quality protein source, as in flakes made directly from wet mixture of whole fish and meats.

Carnivorous fish need one good meal daily. Omnivorous fish can be fed twice daily and herbivorous fish several times. The used seaweed can be used by fishes in case of unpredictable hunger.

Protein Skimmers

This comparatively new tool is useful for use in aquariums in which there is an above average number of fishes or invertebrates and where, for practical reasons, it is not possible to make the frequent changes of water which this situation normally calls for. Protein skimmers are especially useful for the marine aquarist who will be confronted with an excess of protein at more frequent intervals than the freshwater specialist. Proteins, usually in the form of amino acids, are produced by the aquarium animals and will pollute the tank if they are not removed periodically. Protein skimmers fractionate or collect these proteins by circulating the water through a column of vigorously bubbling air; the action of the oxygen in the air on the amino acids produces less soluble compounds which can then be skimmed off the top of the column. To be really efficient, the column of water inside the protein skimmer should be at least 2 metres in height as in professional aquaculture installations, but those available to the amateur aquarist are only 12 to 15 inches (30 to 40 cm) high. Some aquarists have solved this problems by building their own skimmers. The usefulness of the small units to the freshwater enthusiast is doubtful but they are better than nothing if a marine system is being maintained and they do allow a reduction in a frequency of water changes. If you do decide to build your own unit it is important to ensure that the size of the bubbles is as small as possible; the smaller the bubbles the greater the volume of air in contact with the water and the greater the degree of oxygenation. Thousands of tiny bubbles will produce a far more effective fractionation than a few large ones.

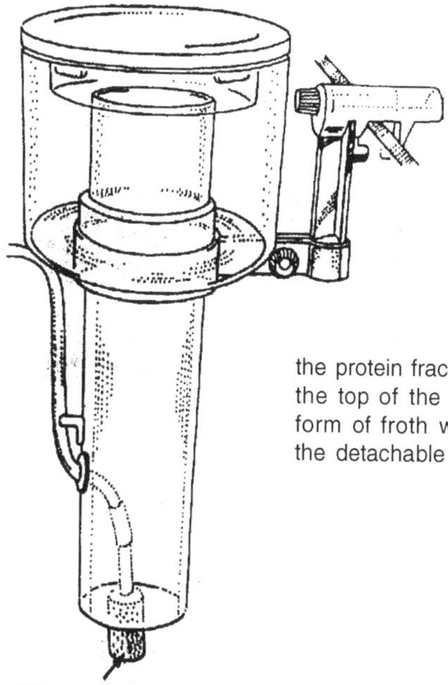

Protein Skimmer

Routine Cleaning

Routine partial changes of the water are most important. The changes are necessary to dilute the build-up of soluble materials. The ideal is to replace 20% of the tank volume weekly. Evaporation losses should be replaced by tap water for salt-water aquaria not by sea water (artificially prepared or natural).

The cleaning can be done by hand; the hands should be washed with soap, after the job and not before because of the danger of introducing soap into the water. Use scraper for algae on glasses.

Algae Cleaners

There are two types of algae remover which have become standard equipment over recent years. The first and less expensive comprises a long rod with a plastic end into which a safety razor blade is inserted. This works exceptionally well and is effective until the razor blade inevitably becomes rusted or blunted by use. The other end of the rod is often modified to act as a planting stick which

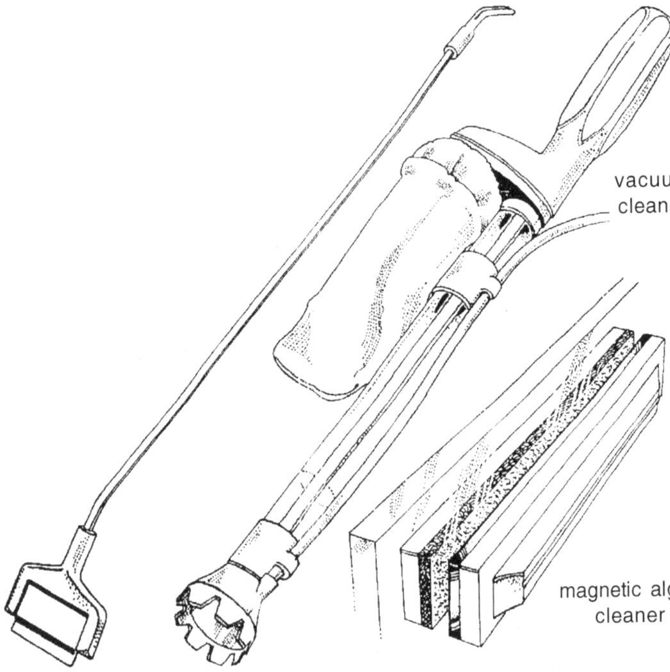

Algae Cleaners

can be very useful especially as plants often become dislodged during the process of removing algae from the sides of a tank. Some models have a plastic blade and, to avoid scratching, these are the only ones which should be used in plastic tanks. A second type of algae cleaner on the market uses two magnets: one bar, which is covered with a rough cloth, is placed inside the tank; the other, covered with a soft cloth and usually bearing some form of a handle, is placed opposite the first magnet but on the outside of the tank and two are held together by magnetic attraction. By firmly, but not too rapidly, moving the outside handle up and down you can effectively remove the algae without wetting your hands or the carpet and at the same time polish the front glass. With a little practice the two magnets can be operated without becoming separated. Powerful magnets are essential and any worthwhile model should be very difficult to separate in the showroom. Those models which come apart easily in the hand will be useless when separated by the glass of the aquarium.

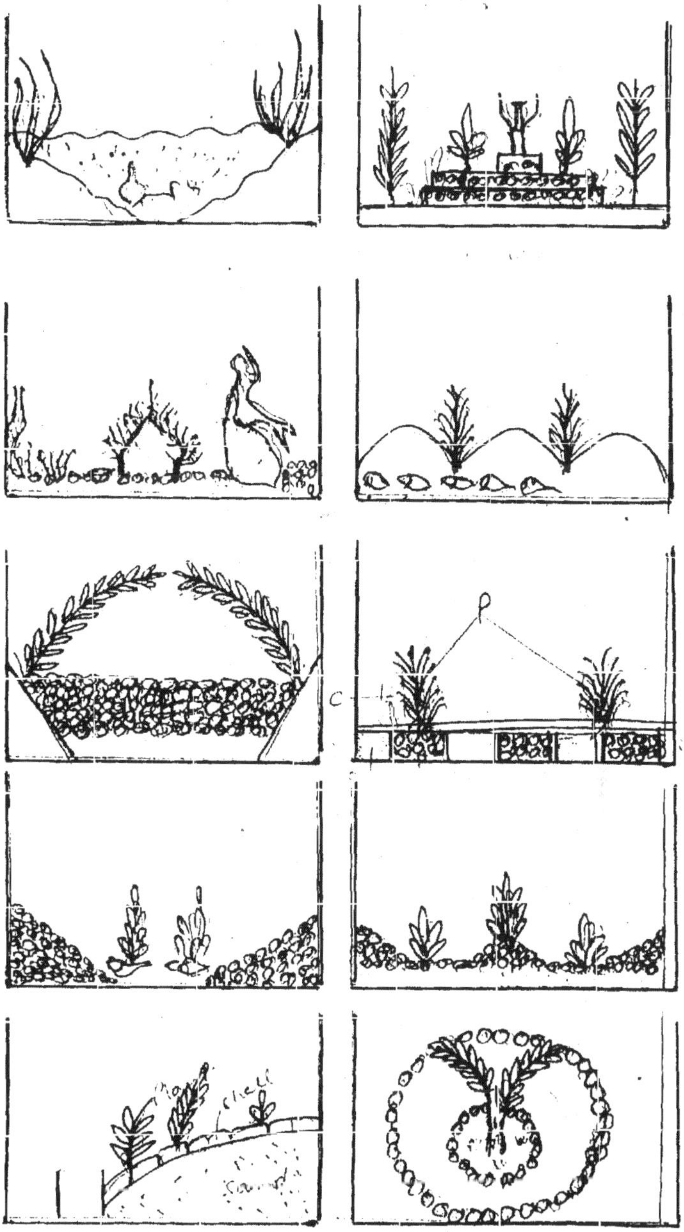

Fig. : Different Decoration Styles of Aquarium

Fig. : Different Decoration Styles of Aquarium

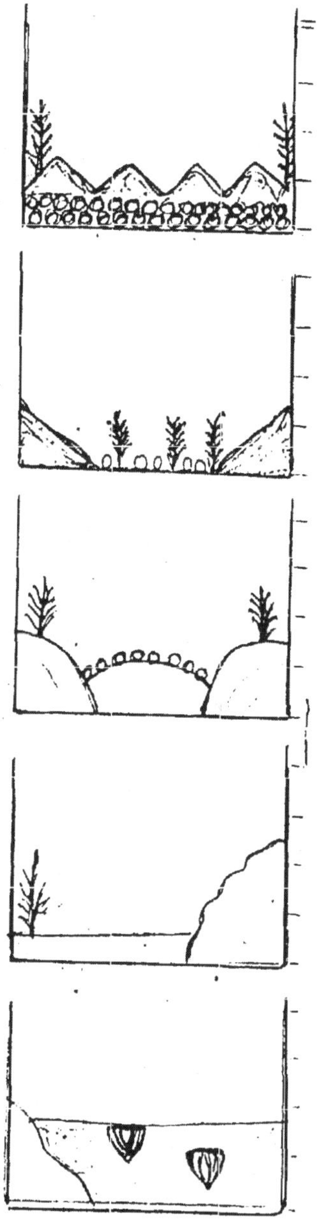

Fig. : Different Decoration Styles of Aquarium

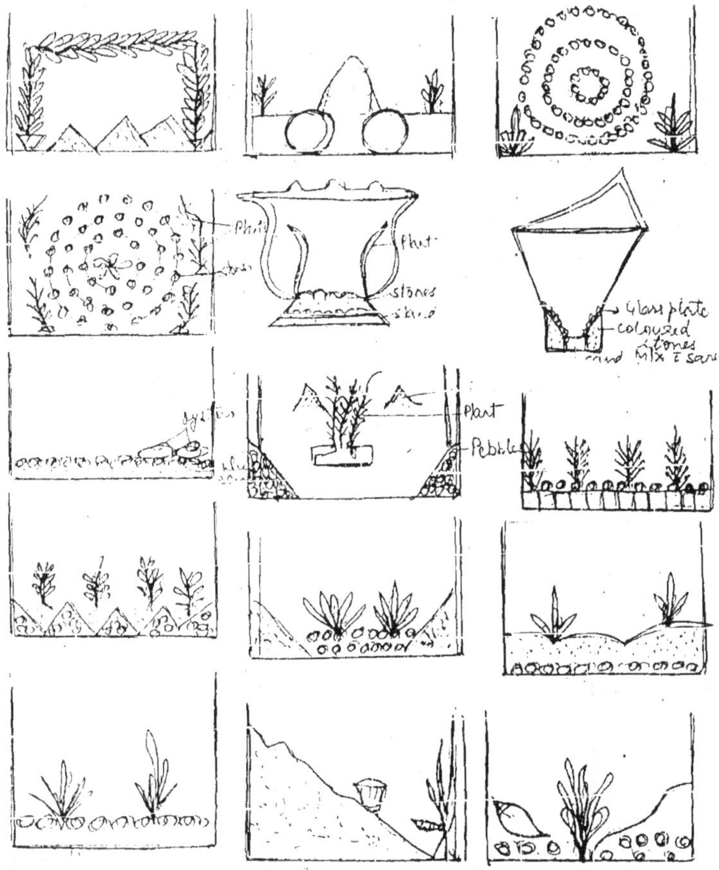

Fig. : Different Decoration Styles of Aquarium

Note: One can decorate both fresh water and marine aquarium according to above mentioned styles.

Marine Species

Fin Fishes

Heterodontus francisci – Californian horn shark
Triakis semifasciata – Leopard shark
Gymnomuraena zebra – Zebra moray
Gymnothorax meleagris – Wide spot moray eel
Muraena pardalis – Dragon face moray
Rhinomuraena amboinensis – Blue ribbon eel
Taeniocongaer digneti – Cortez garden eel
Plotosus anguillaris – Striped catfish eel
Monocentris japonicus – Pinecone fish
Myripristis murdjan – Big eye squirrel fish
Adiryx suborbifalis – Tris sel
Aeliscus strigatus – Striped shrimp fish
Syngnathus leptorhynchus –Bay pipe fish
Doryrhamphus melanopleures – Fantail fish
Hippocampus kuda – Golden seahorse
H. hudsonius – Florida sea horse
Scorpaena guttata – California scorpion fish
Pterois volitans – Lion fish
Synanceia verrucosa – Reef stone fish
Liopoproma rubre – Swissguard basslet
Grammistes sexlineatus – Golden striped grouper
Pseudochromis flanivertex – Orange striped dottyback
Haemulon sexfasciatum – Grayback grunt
Selene vomer – Look down
Equetus lanceolatus – Jackknife fish
Praquest viola – Rock croaker
Gramma loreto – Royal gramma
Platax orbicularis – Round bat fish
Chaetodon faleiferi – Scythe butterfly fish
Chelmon rostratus – Copper banded angel
Holacanthus clarionensis – Clarion angel fish
H. ciliaris – Queen angel fish
Pygoplites diacanthus – Regal angel fish

Labrisomus xanti – Large mouth blenny
Exallias brevis – Short bodied blenny
Lythrypneus dalli – Blue handed goby
Gobiosoma digueti – Banded cleaner goby
G. puncticulatus – Redhead goby
G. multifasciatum – Green band goby
Zanclus cornutus – Moorish idol
Lo vulpinus – Fox face
Balistapus undulatus – Undulated trigger fish
Lactoria farnasin – Five homed cowfish
Diodan holecanthus – Porcupine fish
D. hystrix – Baltoon fish
Bothus mancus – Pacific peacock floundes
Pleuronichthys coenosus – C-oturbot

Invertebrates

Acroposa carnivorous – Staghorn coral
Anthoplausa xanthogrammina – Giant green anemone
Tealia bofoteusis – Strawberry anemone
Radianthus spp. – Clown fish anemone
Cucumarisa miniata – Red cucumber
Toxopneunstes spp. – Sea urchin
Eucidaris tribuloides – Slate pencil urchin
Linckia spp. – Purple sea star
Chossaster papposus – Rose sea star
Cypraea spp. – Cowries
Cyphoma gibbosum – Flaming tongue snail
Murex spp. – Murex
Hermissenda crassicormis – Nudibranch
Octopus bimaculatus – Two spotted octopus
Lysmata wurdemani – Cleaner shrimp
Squilla empusa – Mantis shrimp
Stemopus hispiding – Banded shrimp

Many of the marine fish kept in aquaria are more sensitive to their aquatic environment than the majority of freshwater species. In the marine aquarium, the pH, nitrite and specific gravity levels should be monitored routinely and swiftly corrected if necessary.

Most marine species require a pH of about 8.2, but biological filtration systems tend to produce acidic by-products which slowly reduce the pH. Alkaline materials can be added to the tank such as crushed shells, do1omite granules, marble chips with round smooth edges or sodium carbonate. Nitrite levels should be zero in a few days in the maturing period, after an initial crisis level. A value over 0.1 mg/l should be taken as an indication of some problem in the system. This could be blocked by dirty filter beds, a dead fish or invertebrate, overcrowding or inadequate waterchanging. It is advisable to keep a supply of ready made synthetic sea water for complete water changes when a high nitrite reading is obtained, but before a wtaer remove all living organisms and allow to mature water for reintroduction of fauna and flora.

Sea Water Treatment

First, we deal with the treatment of sea water for through-flow systems and the more sophisticated preparatory treatment for still and recirculation systems. We then consider the maintenance of water quality in recirculation systems.

Sea water treatment is the stage of water quality control between extraction from the sea and its use in an aquarium system. The purpose of treatment is to render the sea water suitable for the culture of the required species. The extent of treatment is dependent upon the tolerance of the culture subjects to prevailing sea water quality being defined as the ability of the sea water to support the species cultured/kept in aquarium.

Sea Water Abstraction and Storage

Choice of abstraction site and method of abstraction

Ideally, water should be drawn from a source where natural populations of the desired culture subjects flourish, but this is not always possible. Many establishments draw their water supply from adjacent estuaries or coastal waters. In some cases, off-shore water is shipped and transported overland in tankers to inland establishments and to coastal sites where it may be more satisfactory for certain applications than the local supply. The choice of a suitable site and the more common methods of abstraction are more important, sub-sand abstraction would appear to be the most attractive technique for most applications.

Sea Water Storage

When water of the required quality cannot be continuously obtained, it must be stored. The water supply may benefit from a period of storage which allows solid material to settle out. At Conwy, the laboratory's water supply is drawn from the adjacent estuary. Outdoor concrete storage tanks (450 m^3), which are topped up by about 30°, of their volume at each high tide, help to buffer variations in salinity.

Storage may also be necessary as a back-up water supply in the event of a drastic, long-lasting deterioration in sea water quality. At Conwy, an annual bloom of the colonial alga *Phaeocystis pouchttti* renders water unsuitable for culturing unicellular algae or for rearing bivalve larvae during the period mid-May to July. A 450 m^3 uncovered outdoor concrete storage tank is filled with water of proven suitability during spring tides at the beginning of May. Once blooming of *Phaeocystis* begins, the laboratory sea water supply is drawn from the storage tank. With careful management, this supply will last for 6-8 weeks. Some deterioration in quality becomes detectable towards the end of this period but at this time, it is often still better than water from the estuary.

Improving Water Quality

Removal of large particles

The extent to which particulate material needs to be reduced depends on a number of factors. These include the method used in abstraction from the sea, the prevailing nature and quantity of such matter, the subsequent steps of treatment employed, the type of culture system used and a knowledge of the tolerance or requirement of the culture subject to such matter. The reduction of the particulate content is usually the first stage in sea water treatment. An exception, discussed later, is where foam separation is applied. Generally the latter technique precedes fine filtration. The following is a guide to commonly used filtration methods.

Sub-sand abstraction

A properly installed and maintained sub-sand filtration unit can be a highly cost-effective method for sea water clarification. With such a facility, the further filtration of larger particulate material is unnecessary. Another advantage where long pipelines from the extraction point are unavoidable is the very considerable reduction in the risk of pipe occulsion through the growth of fouling organisms. The major biological disadvantage is that phytoplankton, other micro-organisms and organic detritus which together constitute the food source of filter feeding animals are excluded. For those working with filter feeders, particularly if held in through-flow systems, the problems of cultivating suitable foods are introduced. The water extracted should be directly suitable for any of the procedures for sub-micron filtration, chemical pre-treatment or sea water sterilization. If, however, any pre-filtered sea water is allowed to stand in storage tanks for an appreciable amount of time, filterable material will develop by the normal processes of particle formation in the sea (Riley, 1970) and from air-borne contamination. The development of particles will be enhanced if the storage tanks are subject to wind action or are aerated.

Sedimentation

Where sub-sand abstraction is either impractical or undesirable, many coastal sited establishments pump directly from the sea into storage tanks. Depending upon the periodicity of pumping, tank size and configuration and the rate of water usage, these tanks can fulfil the role of settling tanks. Negatively buoyant material will settle according to the difference in specific gravity between the particles and sea water. Buoyant materials, including the foam which is produced from organic material during pumping, will rise to the surface initially and may need to be removed by skimming. After peaks of phytoplankton blooms in spring at Conwy, crusts of floating decomposing organic matter are periodically removed with manual skimmers, otherwise they can sink and generate anaerobic conditions at the bottom. The rate of particle settlement is enhanced in shallow tanks, but wind-induced turbulence can impede settlement and further contribute to the particulate content of the water. Settlement tanks especially designed for this purpose are used in freshwater treatment plants, and at the mussel station of Texel, Holland. Some are circular; others are long V-shaped channels.

A relatively recent innovation developed primarily for waste water treatment and the chemical industry is the titled plate or tube separator, sometimes referred to as a particle interceptor. Because they operate at a rapid flow rate, they are suitable for filtering water at source. Essentially, water is pumped upwards through parallel tubes or plates arranged to ensure a laminar flow. Close packing of the elements permits the minimum of distance for particles to settle out on to the surfaces. The angle of tilt is adjusted to ensure that the particle aggregates slide against the direction of flow into a sediment trap. These devices will not, however, work efficiently with finely divided solids which are reluctant to move off the inclined surfaces. In this case, a stable floc formed by chemical treatment improves efficiency. Tilted plate separators are being operated by the National Marine Fisheries Service in a closed-system shrimp culture project at Galveston, Texas (Mock *et al.,* 1986). Parker and Simco (1973) also describe the sue of a tilted tube unit in the culture of fresh water fish in closed systems.

Mechanical filters

Meshes and wadding filters in their simplest form, comprise either a mesh screen or a wad of glass or synthetic wool held in a suitable container. They are widely used to remove large particles from recycled water. With proper maintenance, they provide a cheap method of partial water clarification. Mesh screens, usually manufactured of mono-filament nylon or other synthetic materials, are essentially two dimesionals filters, and have a limited life between cleaning operations. In-depth "wool" filter media (e.g. Perlon, Nylon or Terylene) are more efficient in removing particles, particularly when partially clogged, and are longer lasting. Clarification efficiency is dependent upon the depth of medium used, its packing and the length of time it has been in operation. Neither meshes nor wadding are of value in clarifying large volumes of incoming water. Wadding can be effective when used as a crude filter for low turnover, open flow systems where more sophisticated methods may be very costly. We use nylon mesh filter bags of 45 μm or 61 μm aperture to remove filter-feeding organisms, but not their food, from the incoming water supply to recirculation systems for rearing juvenile bivalves. These systems have a complete water change two or three times each week. Crude filters are rarely built in to pipelines because they need frequent cleaning or replacement.

Sand and diatomite filters are commonly used for water clarification. both being particularly useful because they ran be used online with other water quality control equipment such as charcoal filters, foam separators, ultraviolet light treatment or sub-micron filters. They can cope with small or large batches of water or can be used in continuous operation. As well as the filtration of water at source, they are often used to control the particulate content in recirculation systems either by intermittent or continuous operation. Among their advantages are the use of chemically inert, insoluble filter media, the lack of moving parts and their all-plastic construction.

Sand filters are classified either as slow or rapid. Slow filters are unpressurized and water dispersed over the surface of a bed of clean graded sand and gravel percolates through the media to a discharge pipe. The filters are either constructed as discrete units or they can be built into aquarium tanks to provide *in situ* sub-gravel filtration. Layers of coarse sand and gravel act as a support for a thinner layer of fine sand and ensure an even water flow. It is important that the different layers be well delineated and evenly packed. Uneven particle distribution and packing reduces both the capacity of the filter and its efficiency and can draw suspended material into the bottom layers of the media. The capacity of the filter is increased more by providing a greater surface area or bed than by increasing its depth because it is the upper layer of fine sand which is the site of filtration. Here, suspended particles which bridge adjacent media granules are trapped.

As the layer of fine sand gradually becomes blocked, the efficiency of particle removal improves. Eventually, with continuous use, the rate of water flow drops below acceptable limits and there is a danger that the dissolved oxygen content within the filter bed will fall below that the necessary to support the developed biomass of aerobic micro-organisms. There are of considerable importance in degrading organic material trapped within the filter bed. When clogged, the top layer of fine sand is gently stirred and the turbid over-laver of water siphoned off. Kinne (1976) recommends that the top layer should be replaced once or twice each year with fresh graded sand. Back flushing should be avoided as this, leads to uneven particle distribution and packing.

Slow filters are most commonly used in recirculation systems because of their biological action. They are too slow to be of value in filtering water at source. The water flowing through a rapid sand filter is usually propelled by pressure, the filter media being encased within a robust housing. Being operated in this way, rapid filters provide a much greater throughput of water per unit area of filter bed than slow sand filters.

In managing aquarium seawater, the consistency an convenience by which data are acquired are more important than pinpoint accuracy. The NBS buffer standards give highly reproducible results in seawater (Hawley and Pykowicz 1973, Pytkowicz et al. 1975) and it does not matter that the numbers fall outside the framework of thermodynamics. In determining the acid-base status of aquarium water, the value measured with a pH meter and dilute buffer standards is still "pH" for practical purposes. Further evidence that dilute buffer standards give excellent results can be found in Pytkowicz et al. (1966). They showed that pH measurements made potentiometrically with NBS standards were reproducible to ±0.006 pH units when a common reference electrode was used with different glass electrodes, and that reproducible values of ±0.003 pH units were possible with a single electrode-reference pair.

Bates (1973) recommended that pH meters be standardized at two pH values that bracket the pH range to be tested. The normal pH of seawater is considered to be 8.2, and the instrument should therefore be standardized with NBS buffers 7 and 10.

Salinity (Hydrometer Method)

Salinity can be measured with hydrometers (as specific gravity and converted), refractometers, or by conductivity with salinometers. Precise temperature control is essential when refractometers are used (±0.1°C), which is not possible with hand-held models. Temperature can be controlled more easily in table models. The durability of most conductivity salinometers is less than desirable. The conductor in the leads often breaks with extensive use, resulting in unreliable readings. The hydrometer method given here is inexpensive and accurate enough for aquarium seawater maintenance.

The conversion of specific gravity to salinity is accomplished in two steps: (1) determination of the actual specific gravity with a hydrometer, and correction of the figure obtained to the density in grams per millilitre at 15°C; and (2) conversion of density to salinity by using the, value obtained in the first step.

Procedure

1. Fill a 500 ml graduated cylinder about two-thirds full with sample water.

2. Measure the actual specific gravity and temperature simultaneously. Estimate the specific gravity to the fourth decimal place.
3. Correct the temperature by reading from the values given in Table 4. For example, if the actual specific gravity is 1.022 at a temperature of 20°C, from the table 1.0220 + 11 = 1.0231.
4. From Table 5, read the value for salinity. In this case, the salinity is 31.2 o/oo.

Dissolved Oxygen (Titration Method)

The procedure given here is a titration method for dissolved oxygen adapted from American Public Health Association et al. (1976). Duval et al. (1974) gave a method for the spectrophotometric measurement of oxygen in natural waters. Dissolved oxygen meters are available from several manufacturers, and are reliable to various degrees. The remote stirrers tend to fail on some units; the stirrer blades on others corrode after several weeks of use with seawater. The National Oceanographic Instrumentation Centre* has evaluated several commercially available dissolved oxygen meters and will mail the published results on request.

Sampling

Collect samples in 300 ml glass-stoppered bottles. Siphon water into the bottles with a length of airline tubing, keeping the discharge end of the siphon tube completely submerged. Allow water from the aquarium to overflow the bottle 3 to 4 times its volume. Insert the stopper without trapping air. Analyze within 2 hr.

Reagents

Manganous sulfate solution. Dissolve any of the following manganous sulfates (hydrous) in a little distilled water, then filter and dilute to 100 ml: 48 g MnSO4.4H2O, 40 g MnSO4.2H2O, or 36.4 g MnSO4.H2O.

Alkali-iodide-azide solution. Dissolve 50 g of NaOH (or 70 g of KOH) and 15 g of KI in distilled water and dilute to 100 ml. Add 1 g of $NaNO_3$ dissolved in 4 ml of distilled water.

Starch solution. Blend 0.5 to 0.6 g of soluble starch in a beaker with a little distilled water. Add to a beaker containing 100 ml of

---

* National Oceanogaphic Instrumentation Centre, National Ocean Survey, NOAA, Rockville, MD 20852.

boiling water and continue boiling for a few minutes. Add a few drops of chloroform to preserve.

Sodium thiosulfate stock solution. Dissolve 2.48 g of $Na_2S_2O_3.5H_2O$ in boiled and cooled with distilled water. Dilute to 100 ml with distilled water and add 0.5 ml of chloroform to preserve.

Standard sodium thiosulfate titrant (0.025 N). Dilute 25 ml of sodium thiosulfate stock solution to 100 ml with distilled water. Add 0.5 ml of chloroform to preserve. 1 ml of titrant = 0.2 mg 0.2 ml-1.

Sulfuric acid. Concentrated (*36 N*).

Potassium dichromate stock solution (0.025 N). Dry some $K_2Cr_2O_7$ for 2 hr in a desiccator at 103°C. Weigh out 0.0225 g and dissolve in 100 ml of distilled water in a volumetric flask.

Standardization

1. Dissolve 2 g of KI in 100 to 150 ml of distilled water. Add 10 ml of a solution containing 1 part sulfuric acid and 9 parts distilled water.
2. Add 20 ml of potassium dichromate stock solution.
3. Dilute to 400 ml with distilled water and titrate with the thiosulfate titrant. Add starch solution (1 to 2 ml) near the end point of the titration when a pale yellow color is reached. 20 ml of titrant are required when the standard sodium thiosulfate titrant is exactly 0.025 N.

Procedure

1. Add 2 ml of manganous sulfate solution, followed by 2 ml of alkali-iodide-amide solution. Keep the end of the pipette submerged in the water sample.
2. Replace the stopper without trapping air. Invert the bottle several times, allowing the precipitate to settle halfway each time before inverting again.

   Note: In brackish water and seawater samples, wait 10 minutes before proceeding to the next step.
3. Add 2 ml of sulfuric acid to the sample by letting it run out of the pipette down the inside of the bottle neck. Restopped and invert gently several times until the precipitate dissolves.
4. Pour 200 ml of sample into a beaker. Titrate with sodium thiosulfate titrant to a pale yellow color. Add 1 to 2 ml of

Table 4: Differences to Convert Hydrometer Readings at Any Temperature to Density

Observed Reading	Temperature of Water in Graduated Cylinder (ºC) -2.0	-1.0	0.0	1.0	2.0	3.0	4.0	5.0	6.0	7.0	8.0	9.0	10.0
0.9960													
0.9970													
0.9980													
0.9990	–1	–2	–3	–4	–5	–5	–6	–6	–6	–6	–6	–5	–5
1.0000	–2	–3	–4	–5	–5	–6	–6	–6	–6	–6	–6	–5	–5
1.0010	–3	–4	–4	–5	–6	–6	–6	–7	–7	–6	–6	–6	–5
1.0020	–3	–4	–5	–6	–6	–7	–7	–7	–7	–7	–6	–6	–5
1.0030	–4	–5	–6	–6	–7	–7	–7	–7	–7	–7	–6	–6	–5
1.0040	–4	–5	–6	–7	–7	–7	–8	–8	–7	–7	–7	–6	–6
1.0050	–5	–6	–6	–7	–8	–8	–8	–8	–8	–7	–7	–6	–6
1.0060	–6	–6	–7	–8	–8	–8	–8	–8	–8	–8	–7	–6	–6
1.0070	–6	–7	–8	–8	–8	–8	–8	–8	–8	–8	–7	–7	–6
1.0080	–7	–8	–8	–9	–9	–9	–9	–9	–8	–8	–7	–7	–6
1.0090	–7	–8	–9	–9	–9	–9	–9	–9	–9	–8	–8	–7	–6
1.0100	–8	–9	–9	–10	–10	–10	–10	–9	–9	–8	–8	–7	–6
1.0110	–9	–9	–10	–10	–10	–10	–10	–10	–9	–9	–8	–7	–6
1.0120	–9	–10	–10	–10	–10	–10	–10	–10	–10	–9	–8	–7	–7
1.0130	–10	–10	–11	–11	–11	–11	–11	–10	–10	–9	–8	–8	–7
1.0140	–10	–11	–11	–11	–11	–11	–11	–11	–10	–10	–9	–8	–7

contd....

Table 4–Contd...

Observed Reading	Temperature of Water in Graduated Cylinder (ºC)												
	-2.0	-1.0	0.0	1.0	2.0	3.0	4.0	5.0	6.0	7.0	8.0	9.0	10.0
1.0150	–11	–11	–12	–12	–12	–12	–11	–11	–10	–10	–9	–8	–7
1.0160	–12	–12	–12	–12	–12	–12	–12	–11	–11	–10	–9	–8	–7
1.0170	–12	–12	–12	–13	–13	–12	–12	–12	–11	–10	–9	–8	–7
1.0180	–13	–13	–13	–13	–13	–13	–12	–12	–11	–10	–9	–8	–7
1.0190	–13	–13	–14	–14	–13	–13	–13	–12	–12	–11	–10	–9	–8
1.0200	–14	–14	–14	–14	–14	–13	–13	–12	–12	–11	–10	–9	–8
1.0210	–14	–14	–14	–14	–14	–14	–13	–13	–12	–11	–10	–9	–8
1.0220	–15	–15	–15	–15	–15	–14	–14	–13	–12	–11	–10	–9	–8
1.0230	–15	–15	–15	–15	–15	–15	–14	–13	–12	–12	–10	–9	–8
1.0240	–16	–16	–16	–16	–15	–15	–14	–14	–13	–12	–11	–10	–8
1.0250	–16	–16	–16	–16	–16	–15	–15	–14	–13	–12	–11	–10	–8
1.0260	–17	–17	–17	–16	–16	–16	–15	–14	–13	–12	–11	–10	–8
1.0270	–18	–17	–17	–17	–17	–16	–15	–14	–14	–12	–11	–10	–9
1.0280	–18	–18	–18	–17	–17	–16	–16	–15	–14	–13	–11	–10	–9
1.0290	–19	–18	–18	–18	–17	–17	–16	–15	–14	–13	–12	–10	–9
1.0300	–19	–19	–19	–18	–18	–17	–16	–15	–14	–13	–12	–10	–9
1.0310	–20	–19	–19	–19	–18	–17	–16	–16	–15	–13	–12	–10	–9

contd....

Table 4–Contd...

Observed Reading	*Temperature of Water in Graduated Cylinder (ºC)* 11.0	12.0	13.0	14.0	15.0	16.0	17.0	18.0	18.5	19.0	19.5	20.0
0.9960												
0.9970												
0.9980							3	4	5	6	7	8
0.9990	–4	–3	–2	–1	0	1	3	4	5	6	7	8
1.0000	–4	–3	–2	–1	0	1	3	4	5	6	7	8
1.0010	–4	–3	–2	–1	0	1	3	4	5	6	7	8
1.0020	–4	–3	–2	–1	0	1	3	4	5	6	7	8
1.0030	–4	–3	–2	–1	0	1	3	4	5	6	7	8
1.0040	–5	–4	–3	–1	0	2	3	5	6	6	7	8
1.0050	–5	–4	–3	–1	0	2	3	5	6	7	8	9
1.0060	–5	–4	–3	–1	0	2	3	5	6	7	8	9
1.0070	–5	–4	–3	–2	0	2	3	5	6	7	8	9
1.0080	–5	–4	–3	–2	0	2	3	5	6	7	8	9
1.0090	–5	–4	–3	–2	0	2	3	5	6	7	8	9
1.0100	–5	–4	–3	–2	0	2	3	5	6	7	8	9
1.0110	–5	–4	–3	–2	0	2	3	5	6	7	8	9
1.0120	–6	–4	–3	–2	0	2	3	5	6	7	8	9
1.0130	–6	–4	–3	–2	0	2	4	5	6	7	8	10
1.0140	–6	–4	–3	–2	0	2	4	5	6	8	9	10

contd....

Table 4–Contd...

Observed Reading	*Temperature of Water in Graduated Cylinder (ºC)* 11.0	12.0	13.0	14.0	15.0	16.0	17.0	18.0	18.5	19.0	19.5	20.0
1.0150	–6	–4	–3	–2	0	2	4	5	6	8	9	10
1.0160	–6	–5	–3	–2	0	2	4	6	7	8	9	10
1.0170	–6	–5	–3	–2	0	2	4	6	7	8	9	10
1.0180	–6	–5	–3	–2	0	2	4	6	7	8	9	10
1.0190	–6	–5	–3	–2	0	2	4	6	7	8	9	10
1.0200	–6	–5	–3	–2	0	2	4	6	7	8	9	10
1.0210	–6	–5	–3	–2	0	2	4	6	7	8	9	10
1.0220	–7	–5	–3	–2	0	2	4	6	7	8	9	11
1.0230	–7	–5	–4	–2	0	2	4	6	7	8	9	11
1.0240	–7	–5	–4	–2	0	2	4	6	7	8	10	11
1.0250	–7	–5	–4	–2	0	2	4	6	7	8	10	11
1.0260	–7	–5	–4	–2	0	2	4	6	7	9	10	11
1.0270	–7	–5	–4	–2	0	2	4	6	7	9	10	11
1.0280	–7	–6	–4	–2	0	2	4	6	8	9	10	11
1.0290	–7	–6	–4	–2	0	2	4	6	8	9	10	11
1.0300	–7	–6	–4	–2	0	2	4	6	8	9	10	12
1.0310	–8	–6	–4	–2	0	2	4					

contd....

Table 4–Contd...

Observed Reading	Temperature of Water in Graduated Cylinder (ºC) 20.5	21.0	21.5	22.0	22.5	23.0	23.5	24.0	24.5	25.0	25.5	26.0	26.5
0.9960											19	20	21
0.9970			10	11	12	14	15	16	17	18	19	20	22
0.9980	9	10	11	12	13	14	15	16	17	18	19	21	22
0.9990	9	10	11	12	13	14	15	16	17	18	20	21	22
1.0000	9	10	11	12	13	14	15	16	17	19	20	21	22
1.0010	9	10	11	12	13	14	15	17	18	19	20	21	23
1.0020	9	10	11	12	13	14	16	17	18	19	20	22	23
1.0030	9	10	11	12	13	15	16	17	18	19	21	22	23
1.0040	9	10	11	12	14	15	16	17	18	20	21	22	23
1.0050	10	11	12	13	14	15	16	17	19	20	21	22	24
1.0060	10	11	12	13	14	15	16	18	19	20	21	23	24
1.0070	10	11	12	13	14	15	17	18	19	20	21	23	24
1.0080	10	11	12	13	14	16	17	18	19	20	22	23	24
1.0090	10	11	12	13	15	16	17	18	19	21	22	23	25
1.0100	10	11	12	14	15	16	17	18	20	21	22	24	25
1.0110	10	12	13	14	15	16	17	19	20	21	22	24	25
1.0120	10	12	13	14	15	16	18	19	20	21	23	24	25
1.0130	11	12	13	14	15	16	18	19	20	22	23	24	26
1.0140	11	12	13	14	15	17	18	19	20	22	23	24	26

contd....

Table 4–Contd...

Observed Reading	Temperature of Water in Graduated Cylinder (ºC) 20.5	21.0	21.5	22.0	22.5	23.0	23.5	24.0	24.5	25.0	25.5	26.0	26.5
1.0150	11	12	13	14	16	17	18	20	21	22	23	25	26
1.0160	11	12	13	14	16	17	18	20	21	22	24	25	26
1.0170	11	12	13	15	16	17	18	20	21	22	24	25	27
1.0180	11	12	14	15	16	17	19	20	21	23	24	25	27
1.0190	11	12	14	15	16	18	19	20	21	23	24	26	27
1.0200	11	13	14	15	16	18	19	20	22	23	24	26	27
1.0210	12	13	14	15	17	18	19	21	22	23	25	26	27
1.0220	12	13	14	15	17	18	19	21	22	23	25	26	28
1.0230	12	13	14	16	17	18	20	21	22	24	25	26	28
1.0240	12	13	14	16	17	18	20	21	22	24	25	27	28
1.0250	12	13	15	16	17	18	20	21	23	24	25	27	28
1.0260	12	13	15	16	17	19	20	22	23	24	26	27	29
1.0270	12	14	15	16	17	19	20	22	23	24	26	27	29
1.0280	12	14	15	16	18	19	20	22	23	25	26	28	29
1.0290	13	14	15	16	18	19	21	22	23				
1.0300	13	14	15	16	18								
1.0310													

contd....

Table 4–Contd...

Observed Reading	Temperature of Water in Graduated Cylinder (ºC) 27.0	27.5	28.0	28.5	29.0	29.5	30.0	30.5	31.0	31.5	32.0	32.5	33.0
0.9960	23	24	25	27	28	29	31	32	34	35	37	38	40
0.9970	23	24	26	27	28	30	31	33	34	36	37	39	40
0.9980	23	25	26	27	29	30	31	33	34	36	38	39	41
0.9990	24	25	26	28	29	30	32	33	35	36	38	39	41
1.0000	24	25	26	28	29	31	32	34	35	37	38	40	41
1.0010	24	25	27	28	30	31	32	34	35	37	39	40	42
1.0020	24	26	27	28	30	31	33	34	36	37	39	41	42
1.0030	25	26	27	29	30	32	33	35	36	38	39	41	42
1.0040	25	26	28	29	30	32	33	35	36	38	40	41	43
1.0050	25	26	28	29	31	32	34	35	37	38	40	42	43
1.0060	25	27	28	30	31	32	34	36	37	39	40	42	44
1.0070	26	27	28	30	31	33	34	36	38	39	41	42	44
1.0080	26	27	29	30	32	33	35	36	38	39	41	43	44
1.0090	26	28	29	30	32	33	35	36	38	40	41	43	45
1.0100	26	28	29	31	32	34	35	37	38	40	42	43	45
1.0110	27	28	30	31	32	34	36	37	39	40	42	44	45
1.0120	27	28	30	31	33	34	36	37	39	41	42	44	46
1.0130	27	29	30	32	33	35	36	38	39	41	43	44	46
1.0140	27	29	30	32	33	35	36	38	40	41	43	45	46

contd....

Table 4–Contd...

Observed Reading	*Temperature of Water in Graduated Cylinder (ºC)* 27.0	27.5	28.0	28.5	29.0	29.5	30.0	30.5	31.0	31.5	32.0	32.5	33.0
1.0150	28	29	31	32	34	35	37	38	40	42	43	45	47
1.0160	28	29	31	32	34	35	37	39	40	42	44	45	47
1.0170	28	30	31	33	34	36	37	39	40	42	44	46	47
1.0180	28	30	31	33	34	36	38	39	41	42	44	46	48
1.0190	29	30	32	33	35	36	38	39	41	43	44	46	48
1.0200	29	30	32	33	35	37	38	40	41	43	45	47	48
1.0210	29	31	32	34	35	37	38	40	42	43	45	47	49
1.0220	29	31	32	34	36	37	39	40	42	44	45	47	49
1.0230	30	31	33	34	36	37	39	41	42	44	46	47	49
1.0240	30	31	33	34	36	37	39	41	42	44	46	48	49
1.0250	30	31	33	35	36	38	39	41	43	44	46	48	50
1.0260	30	32	33	35	37	38	40	41	43	45	46	49	50
1.0270	30	32	34	35	37	38	40						
1.0280	31	32											
1.0290													
1.0300													
1.0310													

Table 5: Corresponding Densities and Salinities*

Density	Salinity	Density	Salinity	Density	Salinity
0.9991	0.0	1.0021	3.8	1.0051	7.7
0.9992	0.0	1.0022	3.9	1.0052	7.9
0.9993	0.2	1.0023	4.1	1.0053	8.0
0.9994	0.3	1.0024	4.2	1.0054	8.1
0.9995	0.4	1.0025	4.3	1.0055	8.2
0.9996	0.6	1.0026	4.5	1.0056	8.4
0.9997	0.7	1.0027	4.6	1.0057	8.5
0.9998	0.8	1.0028	4.7	1.0058	8.6
0.9999	0.9	1.0029	4.8	1.0059	8.8
1.0000	1.1	1.0030	5.0	1.0060	8.9
1.0001	1.2	1.0031	5.1	1.0061	9.0
1.0002	1.3	1.0032	5.2	1.0062	9.2
1.0003	1.5	1.0033	5.4	1.0063	9.3
1.0004	1.6	1.0034	5.5	1.0064	9.4
1.0005	1.7	1.0035	5.6	1.0065	9.6
1.0006	1.9	1.0036	5.8	1.0066	9.7
1.0007	2.0	1.0037	5.9	1.0067	9.8
1.0008	2.1	1.0038	6.0	1.0068	9.9
1.0009	2.2	1.0039	6.2	1.0069	10.1
1.0010	2.4	1.0040	6.3	1.0070	10.2
1.0011	2.5	1.0041	6.4	1.0071	10.3
1.0012	2.6	1.0042	6.6	1.0072	10.5
1.0013	2.8	1.0043	6.7	1.0073	10.6
1.0014	2.9	1.0044	6.8	1.0074	10.7
1.0015	3.0	1.0045	6.9	1.0075	10.8
1.0016	3.2	1.0046	7.1	1.0076	11.0
1.0017	3.3	1.0047	7.2	1.0077	11.1
1.0018	3.4	1.0048	7.3	1.0078	11.2
1.0019	3.5	1.0049	7.5	1.0079	11.4
1.0020	3.7	1.0050	7.6	1.0080	11.5

contd....

Table 5–Contd....

Density	Salinity	Density	Salinity	Density	Salinity
1.0081	11.6	1.0111	15.6	1.0141	19.5
1.0082	11.8	1.0112	15.7	1.0142	19.6
1.0083	11.9	1.0113	15.8	1.0143	19.7
1.0084	12.0	1.0114	16.0	1.0144	19.9
1.0085	12.2	1.0115	16.1	1.0145	20.0
1.0086	12.3	1.0116	16.2	1.0146	20.1
1.0087	12.4	1.0117	16.3	1.0147	20.3
1.0088	12.6	1.0118	16.5	1.0148	20.4
1.0089	12.7	1.0119	16.6	1.0149	20.5
1.0090	12.8	1.0120	16.7	1.0150	20.6
1.0091	12.9	1.0121	16.9	1.0151	20.8
1.0092	13.1	1.0122	17.0	1.0152	20.9
1.0093	13.2	1.0123	17.1	1.0153	21.0
1.0094	13.3	1.0124	17.3	1.0154	21.2
1.0095	13.5	1.0125	17.4	1.0155	21.3
1.0096	13.6	1.0126	17.5	1.0156	21.4
1.0097	13.7	1.0127	17.7	1.0157	21.6
1.0098	13.9	1.0128	17.8	1.0158	21.7
1.0099	14.0	1.0129	17.9	1.0159	21.8
1.0100	14.1	1.0130	18.0	1.0160	22.0
1.0101	14.2	1.0131	18.2	1.0161	22.1
1.0102	14.4	1.0132	18.3	1.0162	22.2
1.0103	14.5	1.0133	18.4	1.0163	22.4
1.0104	14.6	1.0134	18.6	1.0164	22.5
1.0105	14.8	1.0135	18.7	1.0165	22.6
1.0106	14.9	1.0136	18.8	1.0166	22.7
1.0107	15.0	1.0137	19.0	1.0167	22.9
1.0108	15.2	1.0138	19.1	1.0168	23.0
1.0109	15.3	1.0139	19.2	1.0169	23.1
1.0110	15.4	1.0140	19.3	1.0170	23.3

contd....

Table 5–Contd....

Density	*Salinity*	*Density*	*Salinity*	*Density*	*Salinity*
1.0171	23.4	1.0201	27.3	1.0231	31.2
1.0172	23.5	1.0202	27.5	1.0232	31.4
1.0173	23.7	1.0203	27.6	1.0233	31.5
1.0174	23.8	1.0204	27.7	1.0234	31.6
1.0175	23.9	1.0205	27.8	1.0235	31.8
1.0176	24.1	1.0206	28.0	1.0236	31.9
1.0177	24.2	1.0207	28.1	1.0237	32.0
1.0178	24.3	1.0208	28.2	1.0238	32.1
1.0179	24.4	1.0209	28.4	1.0239	32.3
1.0180	24.6	1.0210	28.5	1.0240	32.4
1.0181	24.7	1.0211	28.6	1.0241	32.5
1.0182	24.8	1.0212	28.8	1.0242	32.7
1.0183	25.0	1.0213	28.9	1.0243	32.8
1.0184	25.1	1.0214	29.0	1.0244	32.9
1.0185	25.2	1.0215	29.1	1.0245	33.1
1.0186	25.4	1.0216	29.3	1.0246	33.2
1.0187	25.5	1.0217	29.4	1.0247	33.3
1.0188	25.6	1.0218	29.5	1.0248	33.5
1.0189	25.8	1.0219	29.7	1.0249	33.6
1.0190	25.9	1.0220	29.8	1.0250	33.7
1.0191	26.0	1.0221	29.9	1.0251	33.8
1.0192	26.1	1.0222	30.1	1.0252	34.0
1.0193	26.3	1.0223	30.2	1.0253	34.1
1.0194	26.4	1.0224	30.3	1.0254	34.2
1.0195	26.5	1.0225	30.4	1.0255	34.4
1.0196	26.7	1.0226	30.6	1.0256	34.5
1.0197	26.8	1.0227	30.7	1.0257	34.6
1.0198	26.9	1.0228	30.8	1.0258	34.8
1.0199	27.1	1.0229	31.0	1.0259	34.9
1.0200	27.2	1.0230	31.1	1.0260	35.0

contd....

Table 5–Contd....

Density	Salinity	Density	Salinity
1.0261	35.1	1.0291	39.0
1.0262	35.3	1.0292	39.2
1.0263	35.4	1.0293	39.3
1.0264	35.5	1.0294	39.4
1.0265	35.7	1.0295	39.6
1.0266	35.8	1.0296	39.7
1.0267	35.9	1.0297	39.8
1.0268	36.0	1.0298	39.9
1.0269	36.2	1.0299	40.1
1.0270	36.3	1.0300	40.2
1.0271	36.4	1.0301	40.3
1.0272	36.6	1.0302	40.4
1.0273	36.7	1.0303	40.6
1.0274	36.8	1.0304	40.7
1.0275	37.0	1.0305	40.8
1.0276	37.1	1.0306	41.0
1.0277	37.2	1.0307	41.1
1.0278	37.3	1.0308	41.2
1.0279	37.5	1.0309	41.4
1.0280	37.6	1.0310	41.5
1.0281	37.7	1.0311	41.6
1.0282	37.9	1.0312	41.7
1.0283	38.0	1.0313	41.9
1.0284	38.1	1.0314	42.0
1.0285	38.2	1.0315	42.1
1.0286	38.4	1.0316	42.3
1.0287	38.5	1.0317	42.4
1.0288	38.6	1.0318	42.5
1.0289	38.8	1.0319	42.7
1.0290	38.9	1.0320	42.8

Source : Zerbe and Taylor (1953)

* Density at 15ºC. Salinity in parts per thousand.

Table 6: Dissolved Oxygen (mg O_2 l^{-1}) at Saturation in Freshwater, Brackish Water, and Seawater at Different Temperatures

Temp.	*Chlorinity (%)*										
(°C)	*0*	*2*	*4*	*6*	*8*	*10*	*12*	*14*	*16*	*18*	*20*
1	14.24	13.87	13.54	13.22	12.91	12.59	12.26	11.99	11.70	11.42	11.15
2	13.84	13.50	13.18	12.88	12.56	12.26	11.98	11.69	11.40	11.13	10.86
3	13.45	13.14	12.84	12.55	12.25	11.96	11.68	11.39	11.12	10.85	10.59
4	13.09	12.79	12.51	12.22	11.93	11.65	11.38	11.10	10.83	10.59	10.34
5	12.75	12.45	12.18	11.91	11.63	11.36	11.09	10.83	10.57	10.33	10.10
6	12.44	12.15	11.86	11.60	11.33	11.07	10.82	10.56	10.32	10.09	9.86
7	12.13	11.85	11.58	11.32	11.05	10.82	10.56	10.32	10.07	9.84	9.63
8	11.85	11.56	11.29	11.05	10.80	10.56	10.32	10.07	9.84	9.61	9.40
9	11.56	11.29	11.02	10.77	10.54	10.30	10.07	9.84	9.61	9.40	9.20
10	11.29	11.03	10.77	10.53	10.30	10.07	9.84	9.61	9.40	9.20	9.00
11	11.05	10.77	10.53	10.29	10.07	9.84	9.63	9.41	9.20	9.00	8.80
12	10.80	10.53	10.29	10.06	9.84	9.63	9.41	9.21	9.00	8.80	8.61
13	10.56	10.30	10.07	9.84	9.63	9.41	9.21	9.01	8.81	8.61	8.42
14	10.33	10.07	9.86	9.63	9.41	9.21	9.01	8.81	87.62	8.44	8.25
15	10.10	9.86	9.64	9.43	9.23	9.03	8.83	8.64	8.44	8.27	8.09
16	9.89	9.66	9.44	9.24	9.03	8.84	8.64	8.47	8.28	8.11	7.94
17	9.67	9.46	9.26	9.05	8.85	8.65	8.47	8.30	8.11	7.94	7.78
18	9.47	9.27	9.07	8.87	8.67	8.48	8.31	8.14	7.97	7.79	7.64
19	9.28	9.08	8.80	8.68	8.50	8.31	8.15	7.98	7.08	7.65	7.49
20	9.11	8.90	8.70	8.51	8.32	8.15	7.99	7.84	7.66	7.51	7.36
21	8.93	8.72	8.54	8.35	8.17	7.99	7.84	7.69	7.52	7.38	7.23
22	8.75	8.55	8.38	8.19	8.02	7.85	7.69	7.54	7.39	7.25	7.11
23	8.60	8.40	8.22	8.04	7.87	7.71	7.55	7.41	7.26	7.12	6.99
24	8.44	8.25	8.07	7.89	7.72	7.56	7.42	7.28	7.13	6.99	6.86
25	8.27	8.09	7.92	7.75	7.58	7.44	7.29	7.15	7.01	6.88	6.75
26	8.12	7.94	7.78	7.62	7.45	7.31	7.16	7.03	6.89	6.76	6.63
27	7.98	7.79	7.64	7.49	7.32	7.18	7.03	6.91	6.78	6.65	6.52
28	7.84	7.65	7.51	7.36	7.19	7.06	6.92	6.79	6.66	6.53	6.40
29	7.69	7.52	7.38	7.23	7.08	6.95	6.82	6.68	6.55	6.42	6.29
30	7.56	7.39	7.25	7.12	6.96	6.83	6.70	6.58	6.45	6.32	6.19

Source : Spotte (1979) from data in Murray and Riley (1969)

starch solution and continue titration until the pale blue color disappears. After the starch has been added, avoid going past the endpoint by adding the titrant a drop at a time and gently swirling the beaker after each drop. 1 ml of titrant = 1 mg O2 l-1.

Note: If the amount of titrant used during standardization is not exactly 20 ml (step 2), at the end of the test procedure the amount of dissolved oxygen in the sample can be determined by

$$Mn\,O_2\,l\text{-}1 = \frac{\text{ml of test titrant x 10}}{\text{actual ml of standardization titrant}} \qquad (70)$$

5. Compare the measured value with the concentration of dissolved oxygen at saturation (Table 6) at the proper temperature and salinity.

References

Hawkins, A.D. (1981). Aquarium Systems. Academic Press, London, pp. 151-168.

Anthony Evans (1971). Your Book of Aquaria. Fabek and Faber, London, pp. 54-57.

Wickins, J.F. and Helm, M.M. (1986). Seawater Treatment. Academic Press, London.

Chapter 4
Filters and Plants

Introduction

Filters are used to remove unwanted matter from the aquarium and also as a means of introducing desirable substances into the water. There are so many types of filters available. From the simple bottom filter to the more expensive power filters.

Different types of filter

The Bottom Filter

It is very efficient and inexpensive, like the majority of filters it requires an aeration pump to lift the water and induce a flow through the filter media. The filter box, usually made of plastic and more or less cube shaped rests on the gravel in one corner of the aquarium where it can be camouflaged with rock work or tall plants. Air is pumped into the centre section via a porous stone, the rising bubbles lifts the water from this centre section causing water to be drawn in through corners in the lid, or through vents around the base, the water then passes through the filter medias then to the centre section where it is returned to the aquarium proper. A filter of this type will remove much of mulm and larger particles suspended in the water, it will also remove slight cloudiness due to suspended organic matter.

Undergravel or Biological Filter

It has gravel at the bottom of the tank as the filter media. In principle it consists of a number of plastic tubes with small holes along their length, the tubes are connected in a framework terminating in an air-tight tube, the frame work is buried about 1½ or 2 inches under the gravels. Action of the air tight lid draws the water through the gravel, depositing any organic matter plants

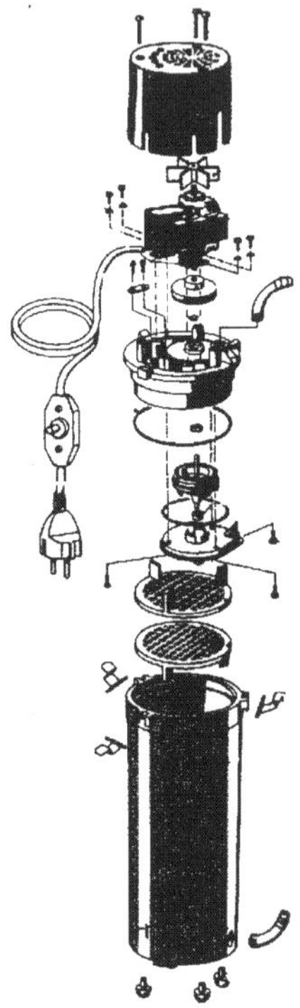

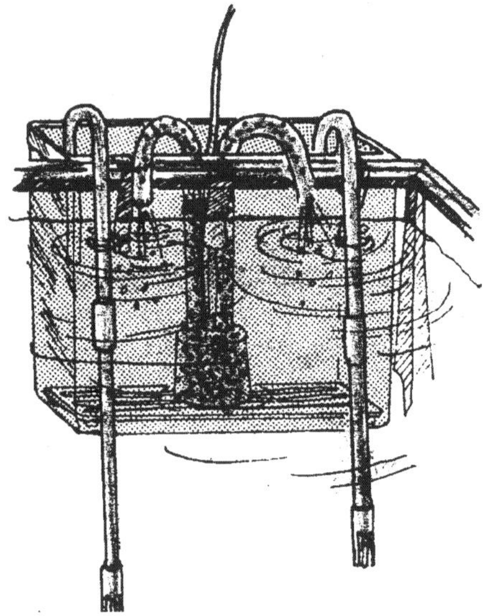

Fig. 2: Bubble up Filter

Fig. 1: An "exploded" view of the Eheim power filter showing the assembly of the many component parts

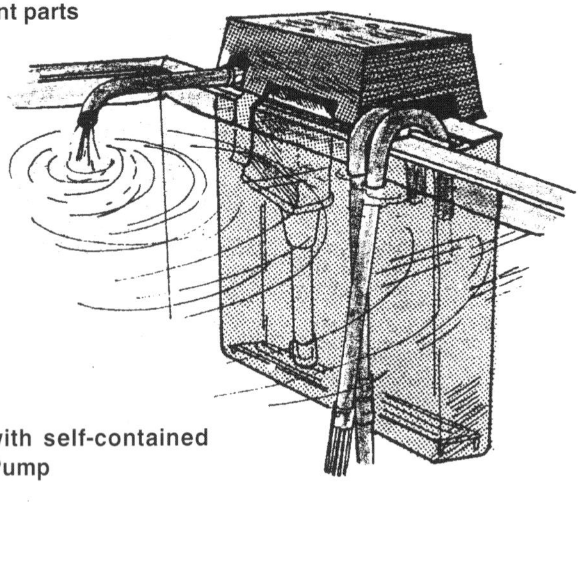

Fig. 3: Filter with self-contained Water Pump

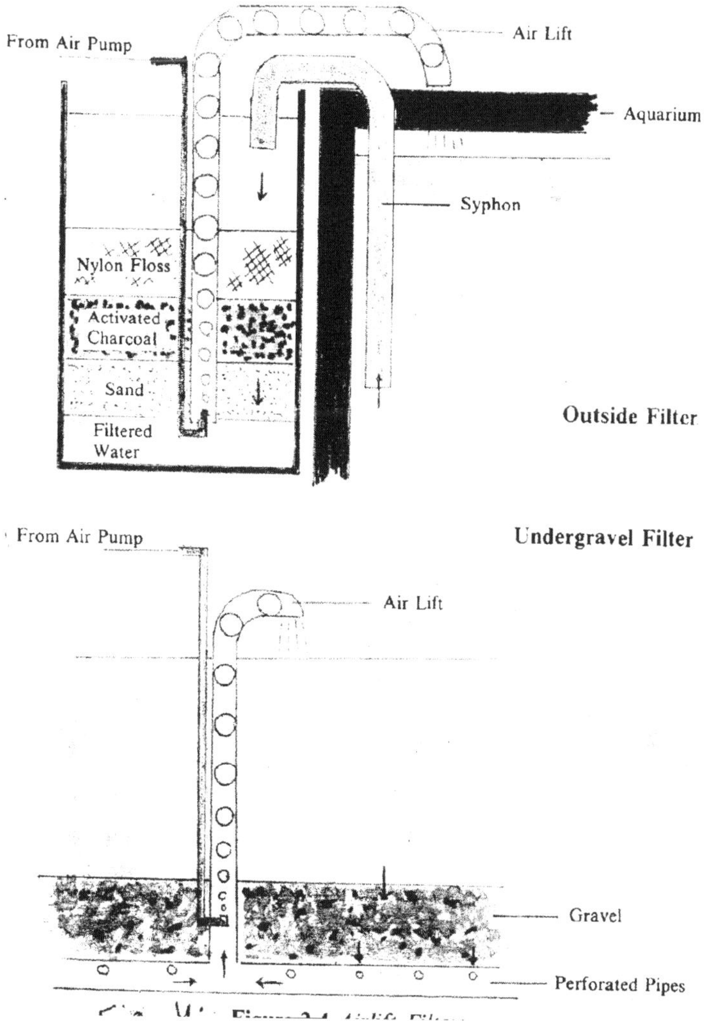

Fig. 4: Airlift Filters

The top diagram illustrates the mechanical type of filtration. The lower illustration depicts a biological type of filter.

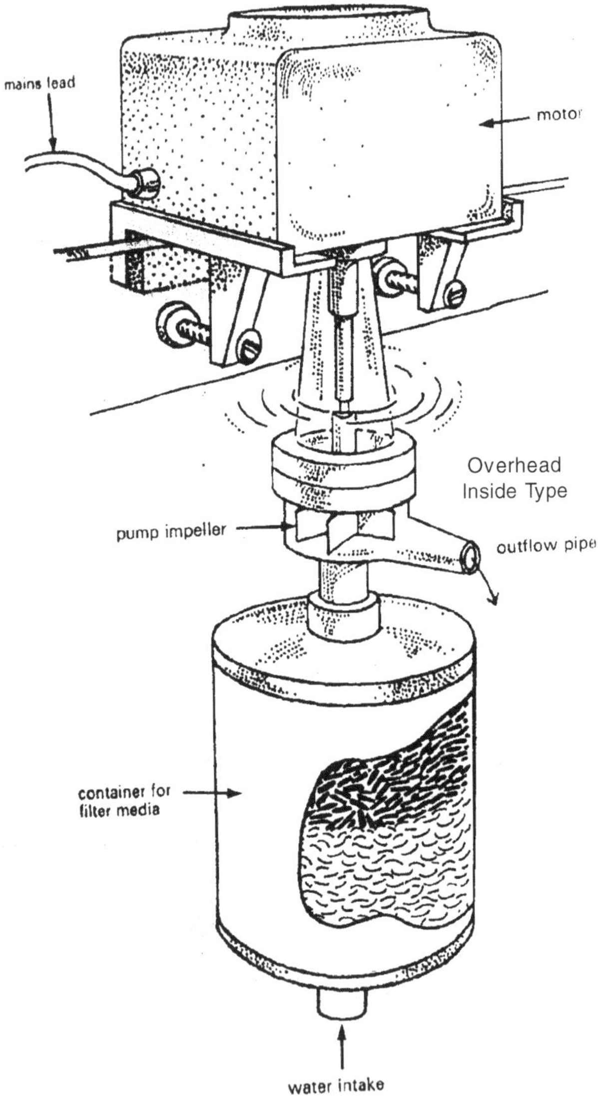

Fig. 5(a): Typical Power Filters

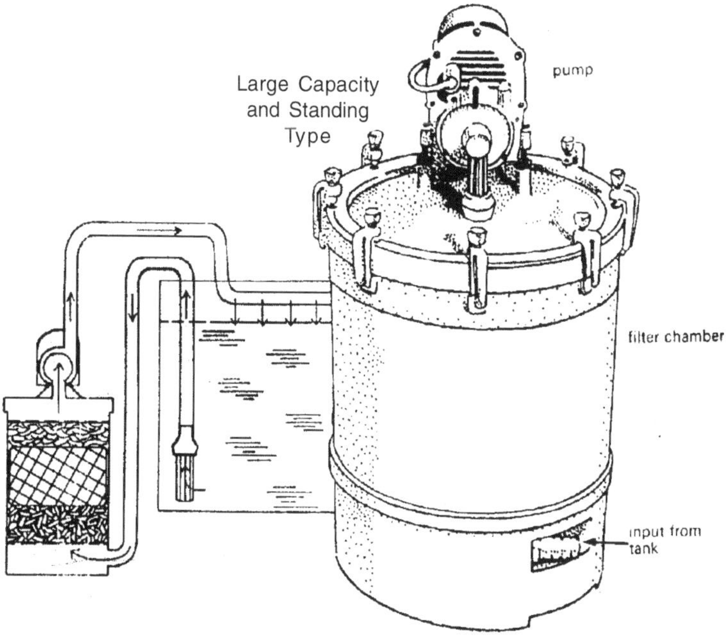

Fig. 5(b): Typical Power Filters

stands to keep the lower areas of the aquarium warmer by transferring the cooler water at the bottom to the top. The biological filter functions by converting waste into minerals and plant nutrients.

Mechanical Filtration

This type of filter consists of a container which is placed in a suitable medium that removes particles from the water as it passes through filter. The simplest type comprises of a small container with a perforated bottom that is positioned inside the aquarium. A quantity of nylon floss or glass wool is placed inside the filter to trap and retain any small particles of matter. A tube leads from a little distance below the filters up, into the filter, media from where the water is discharged above the filter. A smaller tube is connected into this larger tube and is kept air tight : in operation the small tube where it emerges, in the form of bubbles, near the lower ends of the larger tube and raises up that tube. In the process the rising bubbles carry water between them up and this is discharged into the filter from where it quickly seeps through the filter medium and thence back into the aquarium via the perforations. This type of filter is referred to as an "inside filter".

Power Filter

It is undoubtedly the most efficient system but more expensive power filters incorporating high filtration rates and facilities for changing the filter media easily. One type also provides a jet pipe for aeration.

Versatile Filter

It is useful for small tanks, and for experimental work, employs the principle of filer pad, and/or a filter pack compartment, and sieves of various grades from where water can be filtered.

Granule Media Filters

When classified according to flow rate there are 3 types of granular media filters (1) slow (2) rapid (3) high rate. The first two use only sand with a supporting bed of gravel underneath. High rate filters use anthracite and sand (dual media) or anthracite, sand and either granite or multimedia. Mixed media is a synonym for multimedia. High rate filters can be operated at slower rates and rapid sand filters are sometimes designed to accommodate flow rates that approach those of high rate filters. Gravity and presence of filters that are suitable for rapid sand or multimedia filtration.

Gravity Filters

It should be designed with wall to withstand hydrostatic pressure in two direction. It must also be water tight. Minimum wall thickness should be 22.5-30.5 cm.

Pressure Filter

The conventional steel pressure vessel used routinely in waste-water applications are unsuitable for filtering sea water without modification. In short it should be

- Proper surface coating of any metal in contact with seawater.
- Cathodic protection of the metal.

Filter Media

Fine gravel: Usually used to pack the outer layer of a combination filter pack.

Peat moss: Peat moss is used in filter systems to condition the water, especially useful when breeding certain species of fish.

Fiberglass: Used as a fine filter media, but it is not recommended because of the possibility of fine glass splinters entering the tank and lodging in the gills and on the skin of fish.

Nylon wool: Similar to Fiberglass, but much safer.

Napped nylon: A loosely woven material about the thickness of a good quality blanket. Obtainable by the yard and needs only to be cut to size. The same pieces can be used over and over again.

Exchange resins: Use to soften and de-salt the water.

Activated charcoal: Used to purify the water of various organic matter including organic poisons to destroy unwanted material and drugs. Does not remove mineral or inorganic substances.

Plants

Acorus gramineus (Japanese Dwarf Rush)

An attractive plant best suited to the foreground of an aquarium. For the narrow, sword shaped leaves grow only to a few inches height their colour is bright to dark green.

Aponogeton fenestralis (Madagascar lace plant)

The showpiece of all submerged plants. The leaves that are borne on short stems, grow to length of about eighteen inches and to a width of about four inches. It needs a deep rich loam; deep warm (not too cold) water and none too good light. Propagation is by severance of the root stock, and by seeds also.

Cabomba caroliniana (Fanwort)

The well-known *cabomba* of commerce. The slender stems bear coarse-cut, fan-shaped, bright, green leaves. There is a tendency for this plant to become stringy. Nipping the head will sometimes promote bushy growth. It is very beautiful plant that is seen at its best when massed against tall rock work.

Hydrilla verticillata

Another species that closely resembles *Egeria densa*. The alive green leaves are born in whorls intervals up the stem. It is a good oxygenator and easily propagated from cuttings. It grows rapidly in a good light and needs.

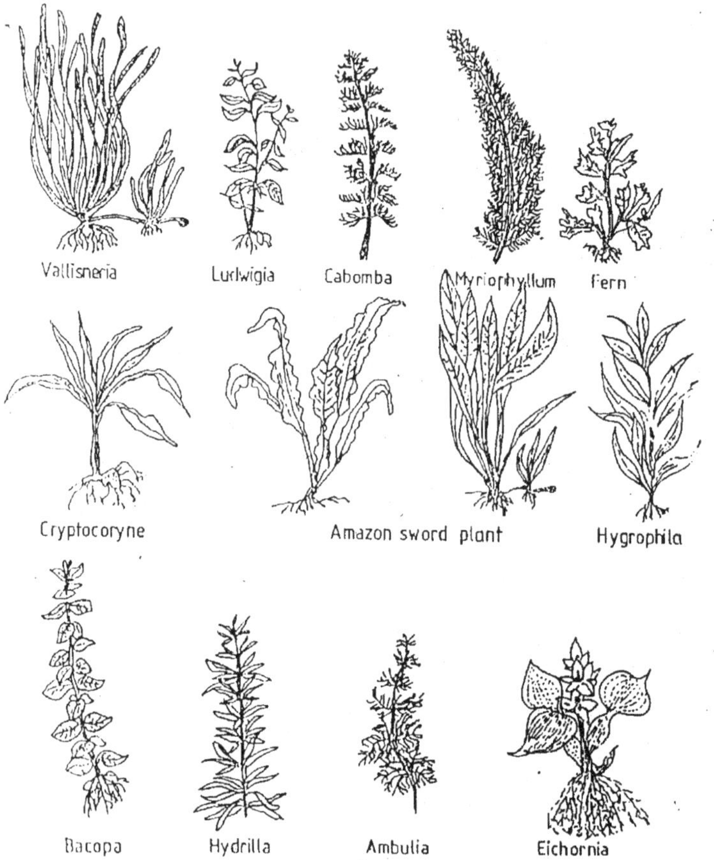

Fig. 6: Aquarium Plants

Lemna (Duck weed)

The dainty little leaves only about 1/3 inch long are dark green and shiny and a quantity of these makes a rather pretty roof to an aquarium in W⁻. It is being used to cut down light. Reproduces by budding resulting in chains of little plants.

Ludwigia natans (Floating lugwigia)

The leaves are lace shape and pointed at both ends. They are joined to the main stem in pairs opposite each other. Their column is brownish green, sometimes brilliant red on the underside.

Fig. 6: A very good design

Fig. 7: Interesting Slopes

Fig. 8: Tank Layout

Fig. 9: Rocks used to support tiers of gravel

Fig. 10: An effective use of twigs etc.

Myriophyllum brasiliense

It is found in south and southern America, it is a deep green colour with distinctly separated feathery leaves forming a shape like a snow crystal star when viewed from above.

Nitella flexis

It is a particularly dainty plant. The leaves, more like small stems from which flowers have been removed, are divided into 2 or more extensions at the tips positioned readily around the main stems. Leaves and stems are dark green.

Vallisneria (Eel grass)

It is an ideal furnishing plant for aquarium with their light green leaves and tall grass like blade propagate by shooting out runners. *V. spiralis* is an excellent oxygenator.

Ceratophyllum demersum (Coarse horn work)

This is strictly a cold water species that does not always thrive in the heated aquarium. Sometime, however, it thrives so well under tropical conditions, and spreads so rapidly the plant has no roots and increases by throwing out fresh branches.

C. submersum is a less common species very similar to *C. demersum* except that the leaves are thinner and more delicate.

Ceratopteris thalictroides (Indian fern)

The species is known by several popular names. The most widely used are Indian fern, Sumatra fern, underwater fern, water sprite. It is a very decorative plant that flourishes in most heated aquaria, especially if the light is subdued. Reproduction is by means of small plants which appear in large numbers on the margins of the mature leaves. The roots are feathery.

Cryptocoryne willisii

This is probably the most popular of the *cryptocorynes*. The leaves are a bright medium green in colour and have wavy edge. Propagation is by means of short runner or by splitting off a part of the parent plant. Once this species has become established an aquarium, they should not be removed unnecessarily as they have an objection to new locations. They grow to about 6 inches in length. The young plants can than be collected and planted where required.

Echinodorus brevipedicellatus (Narrow leaf Amazons sword plant)

This is beautiful plant. The shortish stem are between 2 and 6 inches in length, topped by long, narrow lace-like leaves. In natural waters the plant will grow well above the surface, but in the aquarium they tend to lay flat along the surface. Propogate by runner. Maintained at 21°C temperature.

Echinodorus longistylus **(Long-shafted sword)**

It is a swamp plant. Its shape is typical of the genus with long elliptical leaves about 7 to 9 inches in length on stalks that can reach up to 20 inch high.

Egeria densa **(Argentina anacaris)**

It is a beautiful plant with densely clustered leaves set closely to a central, stem. The roots are sparse and consequently do not anchor very well into the gravel, a little lead may be necessary when first planted, they have the advantage. However, of drawing their nourishment from the water, so they can be cultivated free floating. The column of *E. densa* is a fresh green, and one of its main attraction is that it is quite a fast growing plant. A growth of one inch a day not being unusual, the packed leaves offer a hide out to young fish, but they are too coarse for spawning tropical fish.

Eleocharis acicularis **(Neelde grass)**

The leaves are thin, similar to the shift blades of some grasses and like grass the leaves sprout directly from the root stock, stemless. Dense patches make ideal place for fish spawning, plants with fine needle like blades give excellent protection to eggs scattered by egg laying fish.

These plants have thread like roots which are insufficient to anchor the plant when first planted, therefore, a temporary method of anchoring is necessary.

Pistia stratiotes **(Water lettuce)**

The interesting species of floating plant range over most of the tropical and sub-tropical regions of world. The leaves cascade from a central base like the petals of a flower, are light green in colour, fluted and velvety. Under suitable conditions, the plant will reach a diameter of 4 inches. The root hang down below the plant and are white to blue in colour, they are fine and bushy.

Salvinia auricullata **(Small eared salvina)**

The small hairy oral leaves are attached in opposed pairs to a common stem, and where they join the stem, root like appendages hang down into water. It does not require plenty of light, suitable temperature is 18°C–25°C.

Now you have well acquainted with some plants which can be implanted directly in the sand bed/gravels or in pots or with the help of bamboo sticks or clips in various ways as shown in diagrams. Plants are essential for a complete aquarium ecosystem because it utilizes the faceal matter, yellow damaged leaves, waste feed, undesirable gas materials etc., decorates the aquarium gives of oxygen during daytime by photosynthesis, provides shelter, emergency food, suitable breeding ground, helping in courting to fishes and other inhabitants.

References

Frank W. Orme. Fancy Gold Fish Culture, pp. 40-41.

Stephen Spotte. Seawater Aquariums. pp. 168-188. The Captive Environment.

Hervey, G. F., F.Z.S. and Lack Hems. Fresh Water Tropical Aquarium Fishes, pp. 23-26.

O'Connell, R.F. The Fresh Water Aquarium, pp. 23-67.

Chapter 5

Gold Fish and Anatomy

Introduction

The most valuable fish, based on cost per unit, weight, length are the ornamental fishes. Many of these are caught in their native tropical environment and are shipped to distribution around the world, but as the natural stocks decline, more of these fish are being cultured.

We are not sure how long people have kept ornamental fish, but records show that gold fish were kept in China during the Sung dynasty (starting in the tenth century). Gold fish of different colours and forms were common pets in China by the end of the thirteenth century, spreading to Japan by 1500 A.D., and were brought by sailors to England and France in the seventeenth century. Gold fish were sent to the United States in 1878 and were put on display in Washington, D.C.; the public became excited over the new arrivals, and their relatives (*koi*) are not tropical and may be raised in temperate climates without difficulty.

Commercial culture of tropical ornamental fish began in 1926 in Florida, which is still the centre of the industry. In fact, tropical from that state.

The Gold Fish Varieties

Probably the first-variety to deviate from its wild ancestor, and the progenitor of the many diverse forms of fancy gold fish. It has served to introduce to the wider interests of the fish keeping hobby, being well able to with stand the maltreatment of confined quarters. Glass goldfish – bowl is not the place to keep fish, poor food is given in either too little or too great a quantity, causes dirty water and the consequent many changes of water (Temperature and pH dissolved

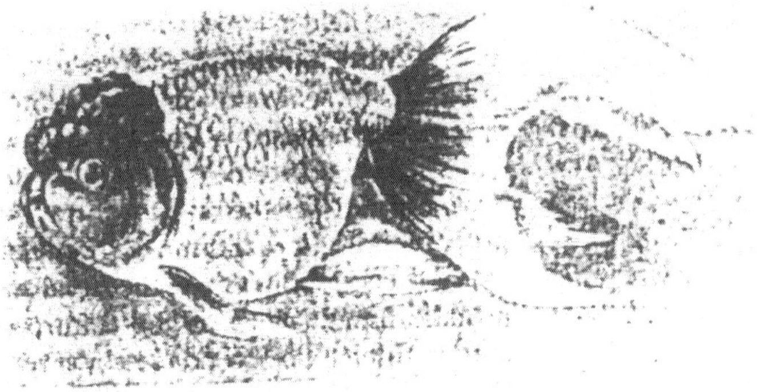

Fig. 1: A "Shukin" Ornamental Goldfish (From Matsubara, 1908)

oxygen that the inexperienced owner to apt to subject it to). Equally at horns in the indoor aquarium or the more rigorous conditions at the outdoor pond, the fish has remained popular with many fish keeper.

Common Gold Fish

A good specimen should have a proper profile, sturdy bright look, with a moderately curved back from the peduncle to the head and a corresponding ventral curve to the underside of the body. The head should be short and wide, with a small mouth, without any trace of snoutiness. The highest point at the back is above the pelvic fins and at this point the dorsal fin commences. The caudal peduncle is short and strong. Pectoral fin shaped and carried stiffly. The eyes are bright and of the normal cris type.

The scales of the common Goldfish have bright metallic in appearance, cover the body in an even patterns of rows, overlapping the root starting at the junction with the knead and running back to the caudal joint. Self coloured reddish orange, orange or yellow one generally preferred to the variegated types, however, red and silver, reddish-orange and yellow or yellow and silver are not uncommon. Black may also be exhibited but this is not a permanent colour and will disappear with time. White *Albino* fish are also available.

London Shubunkin (Nacreous group)

It has extra attraction the as common gold fish – of which it is a variation but lacks the reflective tissue and metallic shine. The transparent scales allow a multiplicity of colours to be seen. Ideally the background colour should be a bright blue intersparsed with patches of red, yellow, brown, violet and black, over which there is black specking. The colour should spread into the fins. Fish that are a single self colour or pinkish white, are not considered to be of any value, suitable for pond and aquarium.

Azumanishiki (Nacreous group)

Identical to the Oranda except for its colours. This and the previous variety are best kept in the aquarium, although they could spend a warm summer in the pond.

Bristol Shubunkin

A slim, streamlined fish, median of its body being less than half its length and the head is less than one-third. The dorsal fish commences on the high point of the back and is almost as high as the body is deep. The pectoral and pelvic fins are long and well developed. The caudal fin is the main feature, being single and well developed with large rounded lobes. The fin should be carried without any sign of dropping. A very attractive fish, probably the most colourful variety, it is also very hardy and does well in both aquarium and pond.

Wakin (Metallic group)

This is the common gold fish of Japan with a similar body shape to the common gold fish and short fins. It has a double caudal fin. Although not as common as the metallic type this is also a nacreous form.

Jikin (The Peacocktail Metallic group)

A very old variety developed from the Wakin. Apart from the caudal fin, the fins and body are the same as the Wakin, although slightly compressed vertically and some thicker in the region of the belly. The best specimens have a silver body with red lips and fins but perfect placement of the red is very rare. The distinguishing feature of this fish is the caudal fin, seen from behind is 'X' shaped and the peduncle is broad. The axis of the caudal fin is almost perpendicular to the axis of the body. Only a very small percentage of the young fish are highly likely to resemble the parents even from the oldest of strains.

Fig. 2: Varieties of tail of Goldfish

1. Common Gold Fish; 2. Peacock tail; 3. Ornada; 4. Calico ornada; 5. Comet; 6. Telescopic eye; 7. Shubunkin; 8. Lion head; 9. Calico; 10. Fringetail

Ryukin (Metallic group)

It is Japan's second most popular variety. In this fish the real deviation is seen towards the short, deep bodied types. The body is short, deepish and moderately compressed often with a pronounced hump at a junction with the head. The fins are longer than these of the Wakin, the caudal fin is forked and divided into two fins; the anal fins are also paired. The Ryukin is popular with professional fish breeders because of its hardiness and ease of management and the high percentage of good progeny, which it produces. Suitable for good pond or aquarium.

Fantail (Metallic and Nacreous group)

The western version of the Ryukin and found in both normal and telescope fish. It is not so deep bodied as the Ryukin and should have no trace of a hump on its back. The dorsal fin rises from the high point of the back to a height of about one-third of the body length. The pectoral, pelvic and dorsal anal fins are short and paddle shaped. The caudal fin is of medium length, moderately forked. The twin fins should be carried swiftly without dropping.

Demekin, Telescope eyed fish (Metallic and Nacreous groups)

Kuro Demkin is velvety black due to an excess of melanin pigment. Some tend to turn red in eye. Aka dem Kin 'is' the red metallic type 'Sanshoko Demakin' is the nacreous form. The eyes protrude greatly from the head. The body shape and finnage is very similar to the Ryukin.

Tosakin (Metallic group)

Tosakin probably a sport from the Ryukin, which it resembles. The main difference are a slightly shallower body and shorter fins together with a peculiarity of the caudal fin. The lower lobes on the caudal fins are, greatly extended with up turned edges; the fin has the appearance of being reversed and spread out in the direction of the head. This variety is raised in Kochi city and its environs near by, where they are kept in very shallow water. Due to the difficulty in swimming they are unable to spawn naturally and must, therefore, be stripped by hand. They are suitable for the aquarium only.

Comet (Metallic group)

A slimmer fish than the Bristol Shubunkin, it has similar finnage except that the caudal fin is as long as the body, and very deeply forked. Yellow is the most usual colour, however, a deep reddish orange is preferred. Suitable for aquarium or pond but probably happiest in the latter environment where it can exercise its occasional turn of speed.

Veiltail (Metallic and Nacreous group)

There are both normal and telescope eyed forms of this variety. The ideal fish has a deep sound body and the depth being greater than half the body length. The height of the dorsal fin should be equal to the body depth and commence on the highest point of the back. The caudal fin is long and broad, falling in greater gradual folds, with a square cut lower edge. The fin is fully divided to form two matching fins lobes. The anal fins are also long and paired. The remaining fins are equally well developed. As the flowing length of the fins makes them susceptible to damage, this fish is best kept in an aquarium to lessen the chance of the fins being torn and where its grace can be fully appreciated.

Oranda (Metallic group)

In all respects similar to the red metallic veiltail, except for the distinguishing feature of the head growth. In Japan this growth is known as the 'wen'. Western Aquarist refer to it as the hood. The head is broad, short and covered by a warty grown of small, round blistery excrescences giving a rasberry like appearance to the head. This as abnormal auricular growth develops in three areas, the cranial infra orbital and opercula. The hood should ideally grow equally in all section, leaving only the eyes and the mouth exposed, however, in this variety, the greatest hood development is usually in the cranial region.

Veiltail oranda

This comes cover in quantities from China, via Hong Kong so that its once prohibitive price has dissolved into mere fractions. A huge cape right over the top of the head and has big fall of tail. It is a very beautiful fish.

Veil tail

Still the most prized and still the most difficult to get. The ideal body of a veil-tail is round, with no trace of oval, really full flanks, even the head and face are round without snoutiness. Their magnificently flowing tail fin is quite twice the length of the body. Gold is the mostly found colour.

Redcap Oranda and Redcap Lion Head (Metallic)

These are colour variation of particular variety. The body is silver and the hood, is restricted to the top of the head. The head colour is red and hence the name. It is suitable for pond or aquarium.

Lion head

It has no dorsal fin and its caudal fin shorter. By contrast, however its lions male growth is enormous quite incredible.

Celestial

This has really the quaintest of fish the body is rounded and plumped in the ideal, no dorsal fin on the smooth line by the back, with a marked caudal peduncle.

Bubble eye

It has two great big pouches under each eye, the whole; eye looks as through it is dropped. The larger the bubble the more expensive the fish.

The Telescopic eyed black moor

It has two forward pointing but short sighted protruding telescopic eyes, a lonely rounded and plump body, with a superb high peaked curved back dorsal fin.

Fan tail

This is a cheap fish body often oval, its dorsal fin is less peaked, less graciously curved.

Anatomy

A fish normally described in four parts, which can be distinguish hed as the head, trunk, tail (peduncle), and fins. The boundary between the head and trunk is determined by the gill opening, and between trunk and tail by the vent.

The skeleton of the goldfish basically consists of a backbone supporting the skull, and providing attachment for the ribs and support for the fins (limbs). The vertebrate of the backnone are so arranged that they allow flexibility of the body, especially in the region of the tail. The first four vertebrate are modified to such an extent that they appear to be almost part of the skull. They have certain curiously shaped elements on each side which provide contact between the ear and the air-bladder, known as the Weberian ossicles. The rear end of the backbone is modified to form the structure of the caudal fin rays.

The skull is made up of a number of bones. The cranium is a box like structure in which the brain is contained, a larger opening at the back allows the spinal cord to pass our backwards. Other apertures allow vigrous network of the nerves to nasal organs and the eyes. Appended from north sides of the cranium, behind the eye openings, is a chain of ones (the suspensorium), and the lower jaw is hinged from the lowest of these bones. The lower law consists of three bones on each side, the upper jaw has four bones – a premaxillary on either side and, behind each, a maxillary. They move forwards when the jaw opens. The mouth, which is roofed by flat bones forming the palatine arch is toothless. Food is masticated by the eight 'teeth' on the pharyngeal bones which bite against a horny pad at the base of the skull, at the entrance to the gullet.

The eye openings are surrounded externally by a ring of circumorbital bones, which contain channels for certain sensory camals. Behind these, hinged to the cramim, are the gill covers-composed of four flat bones that are plainly visible-known as the operculum. The operculum, which protects and cover the gill-cavities, is raised and lowered during the process of breathing.

There are five gill arches, four of which support the gills. The first is the hyoid arch, and consists of large flat bones that form a strong frame for the gill-cavities. The gill arches also have a series of streched 'S' shaped bones (branchiostegal rays) that open like a fan below the operculum to protect the gills. From the junction of the hyoid arches a bone projects, from near the chin, forward into the mouth, supporting what may be called the tongue, whilst another one passes backwards, forming a firm connection with the shoulder-girdle. Each gill-arch comprises a number of slender bones which are suspended in a chain below the cranium, and linked together below. The elements of the last arch are greatly enlarged, and sickle-shaped, barring four rows of strong flattened teeth like processes – these are the pharyngeals.

The shoulder-girdle consists of the pectoral arches, which are attached to the back of the skull on each side, and meet below each arch is made up of several bones joined together, and supports a pectoral fin. Set further back, on the belly, is a pair of pelvic fins, supported by two flat pelvic bones. The pelvic bones are not attached directly to any part of the skeleton.

The fins are supported by rays, each ray being built up from short segments of one, which are very flexible. Each ray of the dorsal, caudal and anal fins are actually double and are formed by elements from each side of the body. Most of the rays are branched, but in some of the fish the first one to three may be spine like and stiff, due to the bone segments being firmly united. The fin rays are hinged upon their supporting bones, to allow movement in one or more directions.

Movable bones are bound together with ligaments and generally cushioned against shock by cartilage.

The stomach, an extensions of the alimentary tract, lies just below the liver and is connected to the coiled intestine, which occupies a position in the lowest part of the body cavity-uncoiled the intestine would be approximately four times the length of the fish and this, in turn, lead to the anus.

The digestion of food is performed in a different manner to that of the higher vertebrates. Fish 'bolt' their food, there is no true chewing or mastication. Food is passed from the mouth to the pharyngeal teeth and the gullet. From the gullet passes through the stomach, where the solid matter is hydrolysed by enzymes in an acid medium.

The blood of the goldfish contains haemoglobin and is, therefore, red in colour. The red blood corpuscles when leaving the heart are deep red (arterial blood), but during the circulatory process, the blood will gather carbon dioxide and become bluish-red (venous blood). The red blood corpuscles in the goldfish are oval in shape and fewer than in the blood of mammals. There are also white amoeboid corpuscles in the blood of fish. The quantity of blood is comparatively small and bears only a small proportion to the body weight-about-one sixty-third.

Respiration in fish is accomplished by means of the gills which are highly vascular structures. The blood is circulated over the branchial lamellae and the blood is aerated. During this process there is a gaseous exchange, whereby oxygen replaces carbon dioxide. The quantity of oxygen utilised by fish is far less than that required by higher vertebrates, so that passive fish, such a goldfish is quite suitable for culture.

The nervous system of the goldfish is a comparatively simple type. The main centre is the brain, with lobes. From the front lobes (the prosencephalon), the olfactory nerves pass forward to the nasal organs. On either side is a large optic lobe, giving rise to the optic nerve which is responsible for sight. Below is the infundibulum, a lobe from which is suspended the important ductless pituitary gland. Behind the optic lobes, on the upper side is the cerebellum, or hind brain, which coordinates muscular activity in response to stimuli received through the senses. Below the cerebellum is the medulla oblongata, which enters the spinal chord where it leaves the cranium.

Locomotion is achieved mainly by flexion of the lateral muscles particularly by the strong muscle in the tail causing the caudal fin to sweep from side to side with a twisting motion.

The reproductive organs consist of ovaries in the female, and testes in the male. The ovaries are yellowish in colour and have a granular texture. The testes have a soft, creamy texture and are much paler in colour. Both ovaries and testes occupy much the same position in the body of the fish, lying just below and behind the air-bladder. They are paired and elongated in shape, and connected to the kidney. As the breeding season approaches and the female become ripe, the ovaries become enlarged, filling the muscular system of the goldfish is not over-complex, consisting mainly of two muscle masses along each side of the central axis. One mass lies above and the other below the axis, and each consists of an upper and lower section. The muscles are in segments known as myotomes, the number of segments corresponding to the number of vertebrate. The dorsal mass is attached to the back of the skull and extends to the full length of the body. The ventral mass is weekly developed at the forward and end shows the strongest development in the region of the tail.

The dorsal and anal fins are controlled by a number of small muscles which expand or contract the fins. The muscle which controls the expansion, or elevation, is stronger. There fore, in a healthy fish, the dorsal and anal fins are seldom folded. The caudal fin is controlled by similar muscles which are also used to move the upper and lower rays from side to side. The front pair of pectoral fins perform a number of movements and thus the controlling muscles are strong and complex. These of the pelvic fins are small and merely being required to spread the rays in fan-fashion like.

The muscles of the skull perform a number of functions, such as operating the gill system, moving the eyes, opening and closing the mouth, these muscles are therefore quite complicated.

The main senses of the goldfish are those of touch, taste, smell, hearing, and a sort of 'sixth-sense' (this 'sixth-sense' is situated in the lateral line). It must be said that there are difference of opinion with regard to the exact nature of these senses, however, it is obvious that they differ from those of land animals which live in a completely different environment.

The organ of smell is not used for respiration, but it does have the faculty of smell, which either attracts or repels a fish when it is near and causes certain odours.

The Auditory Organ

In fish it is much simpler than in man. In fish the external ear and middle ear are non-existent, and inner ear consists only of labyrinth with three semi-circular canals. The vestibule of the labyrinth is dilated into two or more sacs that contain the ear-stones particles, or otoliths for balancing and equilibrium.

Along the side of the body will be found a series of perforated scales known as the lateral line. These scales convert a canal which runs along both sides of the body under the skin, extended from the start of the caudal fin to the head, over the head in three branches, over and under the eye and along the lower jaw. This canal is filled with mucus. But very simply, the lateral line is a sense organ.

Just behind the shoulder girdle lies the liver, a large dark red gland embedded in the liver is the pancreas, its duct enters the intestine close the bile-duct. Near the stomach is the spleen. The kidneys are long and thin, deep red in colour, immediately below the vertebral column. Their duct (the ureter) passes down the back end of the abdominal cavity.

Chapter 6

Ornamental Fishes

One can put any fish in his/her aquarium for demonstration, study and decoration too. All depends upon customer choice and his sense of beauty, asthetic sense. But criteria for selection of fish for home aquaria are — fish should be bright and colourful with extraordinary elegance in morphology. Small in size and compatible with other species. Bold with graceful movements, easy to maintain.

Family – Cichlidae
Suborder – Percoidei
Order – Perciformes
Super Order – Acanthopterygii

Some ornamental fishes of Cichlidae Family are given below.

Aegruidens latifrons **(Blue Acara)**

Also known as the blue spot Cichlid, this species from Panama and Columbia, is one of the best known Cichlids. Fully grown specimen attain a length of 6 inches. But they are mature when only about 3 inches long.

The body colour is mainly a greenish blue, with darker bands vertically positioned along the sides.

Cichlasoma meeki **(Fire mouth)**

The fire mouth is an out standing beautiful fish when in breeding colour. The colour of body is blue green originating in Guatemala and S. Mexico, and growing to about 4 inches.

Cichlosoma nigrofasciatum **(zebra cichlid)**

Originating in central America, this 4 inches cichlid is highly strong but nervous and will dash about the tank if frightened.

Fig. 1: ***Cichlasoma meeki*** **(Cichlidae), Firemouth**

Fig. 2: ***Cichlasoma nigrofasciatum*** **(Cichlidae), Zebra Cichlid**

Fig. 3: ***Etroplus suratensis*** **(Cichlidae), Striped Chromide**

Like most nervous fishes colouring is volatile, but generally the colour is yellowish white with toe silver grey.

Etroplus suratensis **(striped chromide)**

Ranging from the salt and brackish water along the coasts of India and ceylon, this will grow upto 16 inches. Colour change with maturity and age.

Fig. 4: ***Haplochromis multicolor*** **(Cichlidae), Egyptian Mouthbreeder**

Fig. 5: ***Nannacara anomala*** **(Cichlidae), Golden Eyed Dwarf Cichlid**

This species is only really suited for large marine aquaria, it rarely thrives well in fresh water.

Haplochromis multicolor (Egyptian mouth breeder)

This species from Egypt is basically metallic green. The gill corner marked with a blue-black spot.

These fish can grow upto 6 inches in length but aquarium specimen are usually smaller.

Nannacara anomala (Golden Eyed Dwarf Cichlid)

Originating in British Guiena the males of 2¾ inches. So called dwarf are basically green to grassy in colour, with a dark brown spot on each scale and a blue dorsal fin.

Pterophyllum scalare Angel fish

The Angel fish from the waters of the Amazon, Africa and Guiena, are well-known, the long dorsal and anal fins spread gracefully like the wings of a bird, and their slow elegant motions when swimming are so individualistic that the description could fit no other fresh water tropical fish. The silvery body, reflecting tints of blue, has many small speckles not unlike freckles. Evenly spaced black vertical bar's stripe, the body the first one through the eye, the last across the caudal peduncle. One of the body stripes continues into fins. Fins ended up into hair like marginal projection. Occasionally Angel fish will lose their appetite. When this happens try a change of diet, or a partial change of water. Angel fish can grow to about 5 inches (12.7cm) long. They require a temperature range between 75°F (24°C) and 82°F (28°C) and slightly acidic water. Breeding in its natural habitat *Pterophyllum* deposits its egg on reeds. Therefore, the aquarium should be furnished with large leave plants, as Giant Vallisnaria (*Vallisnaria, gigantea*) and Amazon sword plant (*Echinodorus, intermedius*). Breed still, put bamboo, canes or glass rods painted green secured in a vertical position in aquarium at the temperature of about 80°F to 85°F. The eggs hatch in about four days.

Pterophyllum timekei

Grow up to 12 cm. Black and grey with black stripes and albino are common colours, release adhesive eggs, male exhibit parental care.

Family – Anabantidae
Suborder – Anabantoidei
Order – Perciformes
Super Order – Acanthopterygii

***Anabas testudineus* (Climbing perch)**

The Climbing perch has wide natural range including India, the Malay region, Burma, Philippines, Ceylon. The body colour is rich brown or olive green and a bubble nest builders in breeding habit.

***Belontia signata* (comb tail)**

Originating in Ceylon, the comb tail is so named because the rays of tail extend well past the web. The first rays of the ventral fins are also separated.

This is quite a large species growing to about 4¾ inches. It is basically reddish brown marked with indistinct vertical bars of a darker red.

Fig. 6: *Belontia signata* (Anabantidae), Comb Tail

Betta splendens (Siamese fighter)

There is no other aquarium fish quite like the siamese fighter. They are real characters that have a dignity of their own. Almost every aquarist will possess at least one male in a community tank even if he has no intentions of breeding them. If two males are kept in one aquarium they fight upto death.

Betta splendens (Anabantidae)

Siamese Fighter fish size ranges upto 6 cm (2.5 inches) and found in Indonesia and Malaysia, there is no other aquarium fish like siamese fighter fish. The male fighters are the most beautiful, adults develop a most spectacular finnage a high dorsal, large flowing tail and a deep anal fin. Another characteristics is the long and pointed ventral fins which are thrust forward when the fish is excited. When it is in a fighting mood, and this anal fin is stimulated by placing, upward and it holds all fins stiffly erected. "Corn flower" blue and fairy red are the two main colours. But there are many

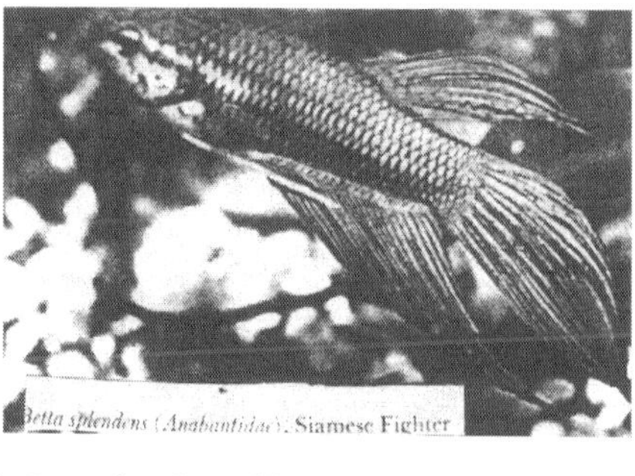

Fig. 7: *Betta splendens* (Anabantidae), Siamese Fighter

colours variations including albino dark blue, green and red all with a metallic quality, the colour completely covering the whole of the body and fins. Females are less colourful and do not possess the layer flowing finnage as of the males. The Siamese fighter enjoys floating vegetation and a temperature range between 78°F (25°C) – 88°F (31°C) and does not require large aquarium. Feeding presents no problems, a varied diet of flake and live food will keep them in excellent conditions. For breeding the bubble nest is built by male, usually under or adjacent to floating leaf. When a male starts building a nest, or even a few bubbles it indicates that it is ready to breed. Female should be introduced a few days before to the male. If they are ready, the male builts nest and proceeds to court the female with height end splash colour and spreads fins. Then female is driven towards the nest and if all goes well, the male will wrap himself around her and fertilize the eggs as they are released. The male will collect the eggs and plant them in nest. Once the spawning is over, the female should be removed. The male will guard the nest, fry hatch in 2 to 3 days once they become free swimming the male must be removed otherwise he will forget all the care and attention, he has lavished on them and eat the lot.

Colisa chuna (Honey dwarf gourami)

Originally from India. It is a pretty fish that grows only to 2½ inches. It's body colour is tan with a yellowish overcast it is marked with a darkish stripe along the side.

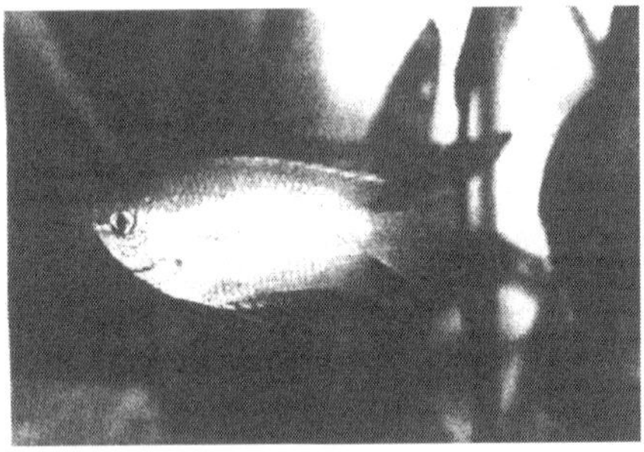

Fig. 8: *Colisa chuna* (Anabantidae), Honey Dwarf Gourami

Colisa labiosa (Thick lipped gourami)

Originally from Burma and attaining a length of 3¼ inches. This is not so thick lipped as its name suggest but it does have a rather square mouth. The body colour is basically silvery green.

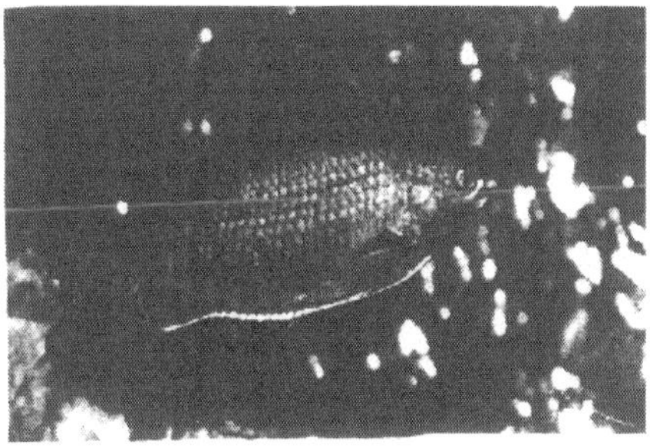

Fig. 9: ***Colisa labiosa*** **(Anabantidae), Thick Lipped Gourami**

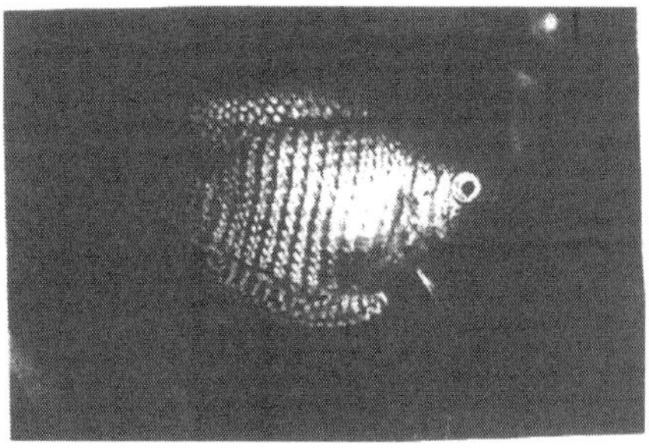

Fig. 10: ***Calisa lalia*** **(Anabantidae), Dwarf Gourami**

Colisa lalia (Dwarf Gourami)

Although the dwarf gourami is smallest of the Gouramis, about 2¾ inches in adulthood, it is by no means the least attractive. Originating in N. India, it has a beautiful colour pattern of red and blue-green bands running vertically across the body. The fins are blue marked with red dots.

Colisa fasciatus Giant gourami

The length of this species is about four and a half inches (11.4 cm). Found in India to Burma. The body colour is warm brown with blue green strips, which slants slightly towards the tail. The edge of the gill cover and belly have brilliant blue green markings. Dorsal fin terminates, and anal fin often fringed with red, and tail flushed red to pink. Because of their larger size it is necessary to provide a roomy aquarium, well planted to provide hiding places for the female. Temperature range 75°F (24°C) to 80°F (27°C). They are not difficult fish to feed, but they do need a varied diet of dried and live foods. For breeding, male builds the bubble nest. When male starts building nest it indicates he is ready to breed. Male does the parental care.

Ctenopoma acutirostre **(Leopard bush fish)**

This species is found naturally in Africa. It is basically a reddish to yellow fish marked with irregular dark spots. Adult grown to about 4 inches. Its body shape is oval.

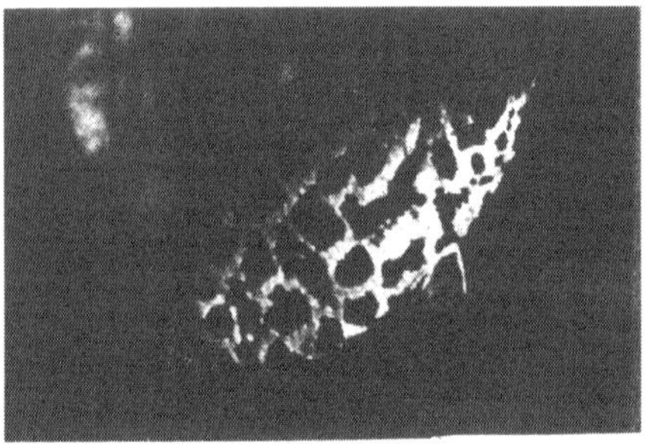

Fig. 11: *Ctenopoma acutirostre* (Anabantidae), Leopard Bush Fish

Helostoma temmincki **(Kissing Gourami)**

Ranging from central India and Indonesia} the Kissing gourmai is an extremely large fish that can grow up to 12 inches and only the small specimens are normally found in home aquaria.

Macropodus opercularis **(Paradise fish)**

The main colouring of this species is a deep rusty brown, marked with emerald stripes. The tail of the male only has long pointed filaments trailing from the tips.

Trichogaster leeri **(Pearl Gourami)**

Also known as lace gouramai. It is a most attractive fish with a body colour of turquoise suffered with a mosaic of pearly dots which extends into fins, and gives the impression of fine lace.

Trichogaster trichopterus

Also called as three spot gournami. Though they have only two spot, eye is considered to be the third spot. Bluish grey colour with two elongated feelers. Male have larger and pointed dorsal fish.

Fig. 12: *Trichogaster leeri* (Anabantidae), Pearl Gourami

Family	**– Characidae**
Suborder	**– Characoidei**
Order	**– Cypriniformes**
Super Order	**– Ostariophysi**

Characins belong to the large group of carp-Iike fish.

Aphyocharax rubripinnis **(Blood fin)**

This small fish from Argentina, grows only to 1¾ inches. It has a silvery body colour. The male, somewhat more slender than the female, has tiny hooks on his anal fin so small that they can not be seen with the naked eye.

Arnoldichthys spilopterus **(striped panchax)**

This attractive little fish from tropical W. Africa, grows to about 2¾ inches. It is silvery with a greenish to bluish sheen, marked rather pretty with a chequered pattern.

Cheirodon axelrodi **(Cardinal tetra)**

This Tetra from the upper Amazon also known as the scarlet

Fig. 13: *Cheirodon axelrodi* (Characidae), Cardinal Tetra

characin. It has a luminous green stripe extending from the eye almost the whole length of the body.

Hemigrammus caudovittatus **(Buenos Aires Tetra)**

As it name suggest, this species is found naturally in the La plateau region around Buenos Aires. It is the largest of the *Hemigrammus* aquarium species, growing nearly 4 inches.

Hemigramus pulcher **(pretty Tetra)**

This is also known as the Garnet Tetra. It is a fish that found naturally in the jungle rivers of the upper Amazon, hence its growing colour. This fish grows to about 2 inches in the aquarium.

Family – Cyrinodontidae
Sub Order – Cyprinodontoidei
Order – Atheriniformes
Super Order – Atherinomorpha

Aphysoemion coeruleum **(Blue Gularis)**

Originating in North Africa. Generally the body is a yellow-brown that darkens on the back and lightens to a blue white on the belly.

Aphyosemion gardneri

Originating in Tropical West Africa. This fish has the typical Torpedo shaped body. The caudal fin has small upper & lower lobes.

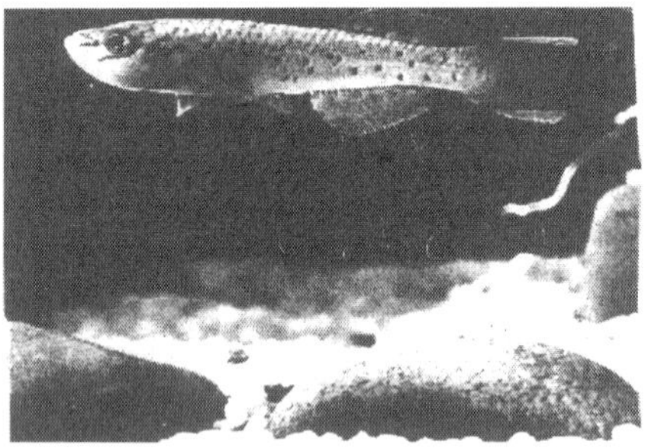

Fig. 14: *Aphyosemion gardneri*

Cyndebias bellottii **(Argentine Pearl fish)**

Originating in La plateau, this attractive species is also known as the Blue chromide. The male have a general body colour of dark safety-blue which changes to blue-green on the fins.

Jordanella floridae **(Flag fish)**

Originating in Florida is a hardy species that can live in quite small aquaria. The name flag fish was bestowed upon this species because of its resemblance to the American flag.

Pachypanchax playfairi **(Play fairs panchax)**

Found naturally in Zimbawe, Seychelles and E. Africa. The body colour of male is yellow to yellow green. The female is similar in colour.

Family – Cyprinidae
Sub Order – Cyprinoidei
Order – Cypriniformes
Super Order – Ostariophysi

This family represents the Barbs and Minnows.

Barbus conchonius **(Rosy barb)**

This species from India, is hardy barbs and has the typical large mirror-like scales which are peculiarly beautiful during breeding time. The main body colour is silver with a rich green back. It is marked with a black-spot on the caudal peduncle. During spawning the male develops a lust of red bronze, but it is a volatile colour that soon melts away when spawning is over. Fully grown Rosy barbs attains a 24 cm in size.

Fig. 15: *Barbus conchonius* (Cyprinidae), Rosy Barb

African Butterfly fish

Pantodon buchholzi size 4 inches, belongs to family Pantodontidae, this single species characterized by unusual pelvic fins with long feelers and large pectoral fin. The latter resemble the butterfly wings, which fish used to jump out of water, but not for gliding. During the day, butterfly fish usually hide under leaves and floating debris. Feed actively only in evening and at night. They feed upon living animals that live at the surface, including terrestrial insects which accidentally land on the water.

***Barbus nigrofasciatus* (Niger barb)**

This species from Ceylon in also known as the black ruby. It is the barb with large light reflecting scales and whilst, it is not particularly brilliant in colour but is very interesting. Adult attain a length of 2½ inches.

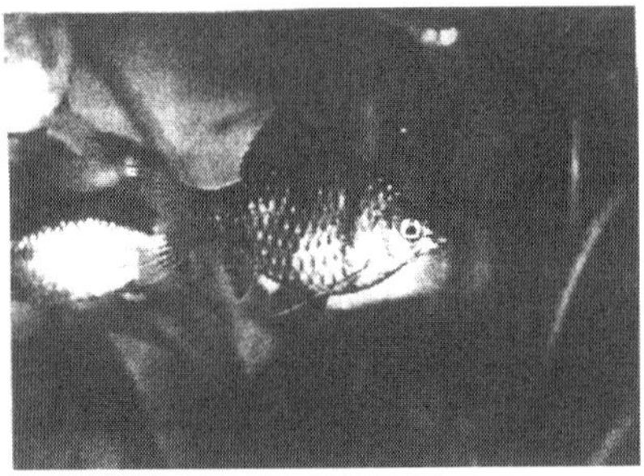

Fig. 16: *Barbus nigrofasciatus* (Cyprinidae), Niger Barb

***Barbus oligolepis* (checker barb)**

Originating in Sumatra, the checker barb is a particularly beautiful fish which is known in America as the beautiful barb. The male is basically red brown darkening along the back, lightening to silver along the belly.

Fig. 17: *Barbus oligolepis* (Cyprinidae), Checker Barb

Barbus tetrazona (Tiger barb)

The general colouring is yellow, striped with four definite black vertical bars, one of which passes through eye. Good specimen have a dark base to the dorsal fin.

Brachydanio rerio (Zebra Danio)

This lively 1¾ inches specimen from Bengal has long been a favourite community fish. It is strictly marked, vivacious hardy and not difficult to breed.

The main body colour is silvery white, marked with a clearly defined deep blue 3-4 horizontal lines that extend into tail and anal fins. Reaches upto 5 cm, can breed when it is 4 cm in length.

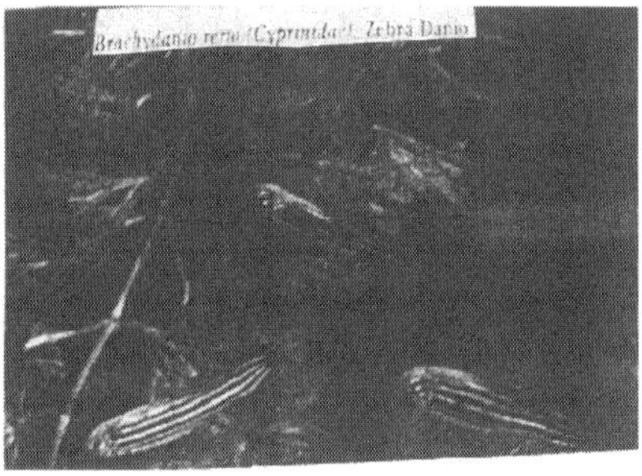

Fig. 18: *Brachydanio rerio* (Cyprinidae), Zebra Danio

Labeo bicolor (Red tailed black sharks)

This species, which can grow up to 6 inches in length, has a definite shark like body, entirely black, except for a fiery red tail. Originating in Thailand.

Fig. 19: *Labeo bicolor* (Cyprinidae), Red Tailed Black Shark

Rasbora maculata **(Dwarf Rashora)**

This species is from Malaya and Indo China. It grows upto 1 inch in length. The body colour is yellowish with a purple red sheen and the belly is greenish to white.

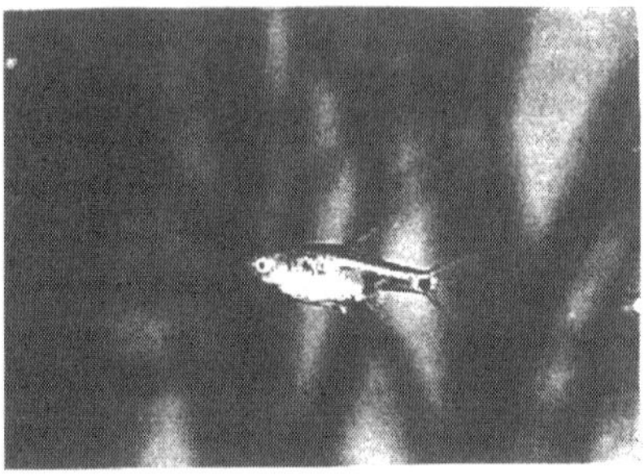

Fig. 20: *Rasbora maculata* (Cyprinidae), Dwarf Rasbora

Family – Atherinidae
Sub Order – Athenrinoidii
Order – Atheriniformes
Super Order – Atherinomopha

This family represents the silver sides, majority of which are found in coastal waters, but a few have taken to fresh water.

Melanotaenia nigrans **(Australian Rainbow)**

The silversides are common saltwater fish, found in nearly all, temperate, water. This is one of the few species that inhabit fresh water prominently. Strangely enough, few of our present day aquarium fish come from Australia, but this peaceful beauty is one exception.

Telmatherina ladigesi **(The celebes rainbow fish)**

This fish as may be expected found naturally in the Celebes island. It is a school fish. They have transparent belly, back and parts of the fins are a pale lemon yellow. They grown upto 2 inches.

Family – Gasteropelecidae
Sub Order – Characidae
Order – Cypriniformes
Super Order – Ostariophysi

***Carnegiella strigata* (Marble Hatchet)**

Also known as striped hatchet fish, this species is found naturally in Guiena and the central and lower Amazon. It has the typical thin blade like deep bellied shape of the hatchet fish. The body colour is silvery with a yellow-gold tint, it is marked with dark brown to black bars across the fore part.

Gasteropelecus levis

Originating in Guiena and Amazon Basin, this species is shiny silver, darkening to olive along the back. It is marked with a narrow strip.

The pectoral fins are well developed into wing like appendages which enable the fish to skim the water species.

Adults attain a length of 2¼ inches.

Fig. 21: *Gasteropelecus levis* (*Gasteropelecidae*), Silver Hatchet

Family – Cobitidae
Sub Order – Cyprinoidei
Order – Cypriniformes
Super Order – Ostariophysi

This family contains the loaches (worm or snake like)

***Botia macracantha* (clown loach)**

Originating in Sumatra and this is golden to orange marked with three vertical bands. The fins are beautifully brick red. It has 4 pairs of barbs. Aquarium species grow to 6 inches.

Fig. 22: *Botia macracantha* (Cobitidae), Clown Loach

Family – Nandidae
Sub Order – Percoidei
Order – Perciformes
Super Order – Acanthopterygii

The Nandis are small fish, perch like and predatory.

Badis badis

The Nandis differs form their related families mainly in skeleton formation. Generally they have large mouth couple of opening to a greater than other members. The majority of Nandis also have almost transparent caudal fins and clear parts in the rear of their anal and dorsal fins.

This originates in India. Adult female are fuller than the male, which retains the hollow under parts:

Family – Poeciliidae
Sub Order – Cyprinodontoidei
Order – Atherinifomes
Super Order – Atherinomorpha

This family represents the live bearing tooth carps. They are found only in the New World.

Heterandria formosa **(Mosquito fish)**

This is one of the smallest fish known, the males are only ¾ inches along, the female are about 1¼ inch long. Because of their small size they are best kept in an aquarium of their own.

The body colour is basically yellow, with dark horizontal stripes, crossed vertically by a few irregular dark bands. The dorsal fin is marked with a red spot.

Live Bearers: Eggs of these fishes are fertilized internally. When fully developed the tiny young ones are released, live by the females. The males are generally small with attractive colours and some possess a special organ known as the gonopodium which develops from the modification of anal fin is used for the transfer of sperms to the female. Some examples are as

Poecilia reticulata **(guppy)**

These are also known as Rainbow fish. The males will grow to about one inch and are the possessors of the bright beautiful colouration. These guppy, are live bearers.

Prolific breeder, produce 20-50 young ones in four-six weeks male are brightly coloured size 3-4 cm.

Poecilia latipinna

Body is uniformly coloured in black that some times even the eye can not be distinguished ripe female can produce 20-50 young ones, size is 8-9 cm.

Xiphophorus helleri

Several colour variations, the lower portion of tail fin of male is generally drawn out into sword like spike, size 8 cm.

Limia caudofasciata **(Blue poecila)**

This species from Japan is a truly beautiful aquarium fish. They are vividly coloured. The colour of the body is generally olive with dark markings along the sides.

Male grow to 1½ inches females to about 2¼ inches.

Limia nigroufasciata **(Hump-back Limia)**

This is found naturally in Haiti where it appears in great numbers.

Adult males are blue-green pale to yellow on the belly, the upper part of body is marked indistinctly with dark vertical bars.

They grow to about 2½ inches. They are active and peaceful fish.

Molliensia latipinna **(Sailfin mollie)**

Originating in North Carolina, Florida and N.E. Mexico, this species obtains its popular name sailfin from the large dorsal fin of the male.

These are large fish growing to 3½ inches in length, the colour of male is bluish overlaid with a reticulated design and horizontal stripes.

Family – Scatophagidae
Sub Order – Percoidei
Order – Perciformes
Super Order – Acanthopterygii

The, 'scats' have laterally composed bodies.

Scatophagus argus **(spotted scat)**

Found naturally in the East India, *S. argus* has a very slim, laterally compressed body, which is almost disc-Iike in out line. It is coloured grey to green, or pink to bronze-yellow, and is marked by large circular spots.

Selenotoca papuensis

This fish has a general body colour of silver marked with seven or eight black vertical bands on the upper part of the body. The body shape is vertically compressed and oral.

Family – Tetradontidae
Sub Order – Tetradontoidei
Order – Tetradontiformes
Super Order – Acanthopterygii

This family contains the 'four toothed' puffers.

Tetraodon fluviatilis **(Green Puffer)**

Found naturally in India, Burma and the Malay Peninsula in fresh and brackish water. Generally body colour is greenish on the dorsal surface and side, paling to whitish, yellow or grey on the underside. It is marked with more or less round dark spot. Growing to about 7 inches. They have the ability to puff themselves into a ball.

Family – Hemidontidae
Sub Order – Characoidei
Order – Cypriniformes
Super Order – Ostariophysi

Nannobrycon eques **(Pencil fish)**

This fish in common with other species of the genus, swim slowly by vibrating motions of the vertical fins and the body, to all intents and purposes, held rigid. This given the fish a wooden appearance hence its name pencil fish. They are peaceful and best kept with their own species or with closely related species.

Nannostomus marginatus **(Dwarf pencil fish)**

The pencil fish are slender, or very slim fish, this particular little species rarely exceeds 1¼ inches. A plastic grill or glass marbles should cover the floor of the breeding aquarium.

Family – Centrarchidae
Sub Order – Percoidei
Order – Perciformes
Super Order – Acanthopterygii

Elassoma evergladei **(Pygmy Sunfish)**

This species is brownish with darker spots and rather indefinite vertical bars. The colour is much enhanced during breeding when the males become blue-black. The species is found naturally in the Ever Gladier Florida, and grows to only 1–1¼ inches.

Mesogonistius chaetodon **(Black-banded sunfish)**

Originating in the U.S.A. this 4 inches species has a silvery body marked with not too well defined vertical bars. The male has a shiny spot on the gill corner. Younger fish are the prettiest, and they are happy in the company of other fish.

Family – Monodactylidae
Sub Order – Percoidei
Order – Perciformes
Super Order – Acanthopterygi

This family comprises one genus-*Monodactylus*.

Monodactylus argenteus

The attraction of this fish is derived from its almost circular laterally compressed silvery body. These have the length of about 5 inches in the aquarium.

Family – Gobiidae
Sub Order – Gobioidei
Order – Perciformes
Supder Order – Acanthopterygii

Family of mostly small fish in which the ventral fins have grown together to form a sucking disc.

Brachygobius xanthozonus **(Dumble bee)**

These tiny 1½ inches. Fish from India and Indo-China certainly gives the appearance of a dumble bee with their yellowish body and dark vertically striped bands of different widths. Males are brighter in colour.

The ventral fins are united to form a sucking disc. Which enable them to cling to rocks, grass etc.

Family – Callichthyidae
Order – Siluriformes
Super Order – Ostariophysi

Unarmoured cat fishes.

Corydoras **agassizi**

This is one of the slightly darker aquarium cat fish, which will attain a length of about 3½ inches. It has the typical appearance of a cat fish, with a particularly flat on the underside and has spade-like head. The fore part of the dorsal fins is dark grey. This is found naturally in Amazon and Western Brazil.

Corydoras aucuatus **(The streamlined Cory)**

Originating in the upper Amazon, the body colour of this fish is silvery white with a greenish blue overall sheen. It is marked with dark curving line from the eye, along the upper part of the body, to the base of the caudal fin. Adult fish attain a length of 2½ inches.

Corydoras julii **(Leopard cat fish)**

Originating in East and North Brazil the Leopard cat fish is quite a popular fish. Its body colouring is a whitish-grey, peppered with black spots these spots extend over the nose.

Fully grown specimen are not very large about 2½ inches long.

Family – Loricariidae
Order – Siluriformes
Super Order – Ostariophysi

Gill sac cat fish possessing an auxiliary breathing organ permitting the fish to leave the water for long periods.

Egg Layers: The fishes release eggs that are seen fertilized externally by males. Some builds nest and lay the eggs in the nest, some will pick up the eggs and incubate them in their buccal cavity: some release adhesive eggs which get attached to different substrata including plant leaves and some eggs sink to the bottom and develop into young ones.

Carassius auratus

They occur in a variety of colours ranging from pure gold to red orange, black and albino. Grow up to 20 cm in length and bred at 6 cm.

Chanda ranga

Grow up to 7 cm length this fish is diamond shaped and the forehead is conspicuously indented, body is translucent, release adhesive eggs.

Gynanocorymbus trenetzi

Short body having a silvery-grey colour and two black vertical bands behind gill cover, size 6-8 cm.

Ococinclus affinis **(Dwarf sucking cat fish)**

This species originates in Riode Janeiro. It is scavenger. It has a sucking organ. It has a charm in stream lined shape of its body.

Another peculiarity of this fish is that it will sometimes swim upside down immediately below the water surface, clinging to it as it apparently sucks in any floating food.

Family – Schilbeidae
Order – Siluriformes
Super Order – Ostariophysi

Cryptoterus bicirrhus **(Glass cat fish)**

The glass catfish is an oddity from Malaya. The body is really transparent. The skeleton is quite easily seen.

It has a long anal fin which occupies most of the ventral surface of the body. Fully grown specimen in the aquarium rarely exceed 2-3 inches.

Some Salt Water Fishes

Warsses (*Coris gaimardi*)

They are among the most conspicuous and abundant of reef fishes. Most are tropical and some species occur in temperate waters. All are marine. They are diurnal. Some burrow in the sand at night for protection. Sex-reversed individuals play a major role in the reproduction of many species and often initiate group spawning. A peaceful aquarium species. The brightly coloured juveniles are prefered as with other species well aerated should be provided at 22-27°C.

Jaw fish (*Opolategnathus auriframs*)

They have large mouth and big eyes. They live in burrows after construction. Dorsal and anal fins are large and continuous with the notch. It is a good aquarium species. Keep in tanks. Provide a bottom of sand minced with mouth size building material and a few larger rocks. Mixed material from coral and beaches works well to keep at temp. 22-27°C.

Gobies (*Nemateleotris magnifica*)

They grow upto Size to 6 cm. and found in Tropical Indo Pacific from east Africa to Piteairn Island and North to Hawaii. Provide aquarium with a substrate of 10 cm or more for Gobies. It will be suitable for the construction of burrows. A mixture of sand, gravel, pebbles, coral rubble and a few small stone usually work, well. Introduce other species only after the gobies are established at 22-28°C.

Puffers (*Arathron meleagris*)

They are soft bodies fishes which can gulp water to enlarge themselves and thus discourage predators. The dorsal and anal fins have soft rayed only and along with anal fin provide the main propulsion.

They produce a toxic mucus but can be kept successfully in large aquaria at 22-28°C regular partial water changes are recommended, especially when this species is first introduced.

Family – Ceratodontidae
Order – Ceratodontiformes

Neocerotodus forsteri (Australian lungfish)

Size: 180 cm.

Range: Queensland, Australia.

Feeding:

In nature: fishes, amphibians, invertebrates and some plant materials, which is probably ingested because of minute organisms that live upon it.

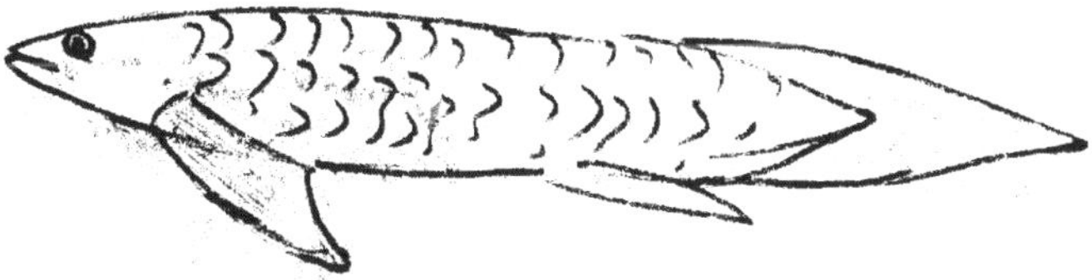

Neocerotodus forsteri **(Australian lungfish)**

The Australian lungfish is considered the most primitive of the living lungfishes since it possesses only a single lung, cannot aestivate and is able to survive without breathing at the surface. This is a robust species characterized by large paddle like fins and large cycloid scales. The Australian lungfish is a protected species and is not available to the aquarists, but it can be seen at some public aquaria including the Vancouver Public Aquarium in Canada.

Family – Mormyridae
Order – Osteoglossiformes

Hyphessobrycon serpae **(Serpa tetra)**

Size: 5 cm

Range: Amazon basin and Guyana

Feeding:

In nature: Insects and other small organisms.

In aquarium: Does well on prepared aquarium foods.

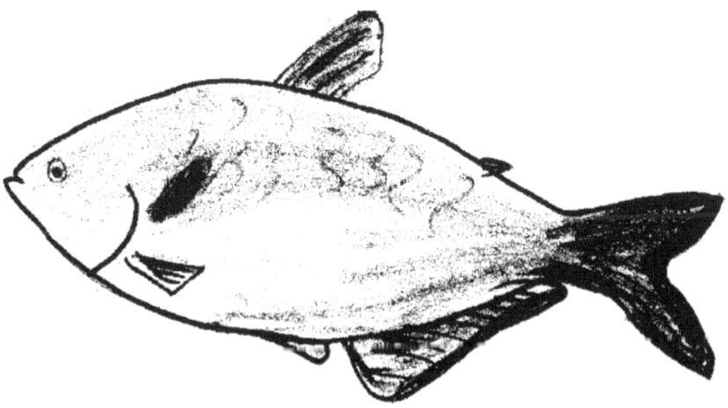

Hyphessobrycon serpae **(Serpa tetra)**

Aquarium care: A hardy species and easy to keep; often recommended for beginners. Water quality is not critical; but should be soft and slightly acid for breeding keep in at least a 30-litre (8 gal) aquarium, planted with fine-leaved plants. Temperature 22-27°C (72-80°F).

Hypgessobycon callistus

Body is pinkish with deep colour fins, very active fish and grow up to 4-5 cm. Eggs are brownish.

Family – Polypteridae
Order – Polypteriformes

***Polypetrus bichir* (Nile bichir)**

Size: 70 cm.

Range: Nile river and adjacent lakes and rivers.

Feeding:

In nature: Insect larvae, worms, small invertebrates and fishes.

In aquarium: Feed a varied diet including worms, mealworms, insect larvae, pieces of meat and fish.

***Polypetrus bichir* (Nile bichir)**

Aquarium care: Bichirs are hardy and often become very tame although they are not always available at tropical fish stores, they are worth looking out for. Provide an aquarium of suitable size with rocks and roots for cover. Water quality is not critical and they do well at temperatures of 20-30°C. Bichirs may appeal to the aquarist who is interested in fishes with distinctive biological characteristics, their ancient origin and unique swim bladders give them this quality.

Family – Characidae
Order – Characiformes

Serrasalmus nattereri (Red piranha)

Size: 30 cm (12 inch)

Range: Amazon and Orinoe river basins.

Feeding:

In nature: Carnivorous with sharp cutting teeth that allow it to feed on large prey, primarily fishes but also other vertebrates including mammals, insects and invertebrates.

In aquarium: Pieces of fish supplemented with meat.

Serrasalmus nattereri **(Red piranha)**

Aquarium care: Require a large tank; water quality is not critical; usually kept between 21 and 27°C (70-80°F). Highly aggressive, it is best to begin with a group of young specimens and rear them together in order to minimize fighting. There are approximately 15 species of piranha in the genus Serrasalmus; many are similar in appearance and difficult to distinguish. *S. nattereti* is most commonly imported but can confused with other species. The black piranha (*S. rhombeus*) is well known to aquarist and sometimes erroneously called *S. niger*.

Family – Characidae
Order – Characiformes

Astyanax mexicanus (Blind Maxican cavefish)

Size: 9 cm (3.5 inch)

Range: Caves of east-central Mexico.

Feeding:

In nature: Available animal matter.

In aquarium: Normal aquarium diets

Astyanax mexicanus **(Blind maxican cavefish)**

Aquarium care: A hardy species, but does best if kept in groups of its own kind in a spacious tank; water quality is not critical; does well over a wide temperature range 18-24°C (64-75°F) is probably optimal. An illuminated tank with a cave-like decor provides realism.

These living in a dark environment are unpigmented and appear pink, when kept in an environment with light they become silvery.

Family – Gasteropelecidae
Order – Characiformes

Carnegiella strigata **(Marbled hatchetfish)**

Size: 4.5 cm (1.75 inch)

Range: Guyana, Surinam, Fresh Guiana, Amazon basin.

Feeding:

In nature: Small insects.

In aquarium: Should have insects to thrive, some prepared food may be adequate.

Carnegiella strigata **(Marbled hatchetfish)**

Aquarium care: Keep at least six in as long an aquarium as feasible. Provide a good filting cover but have an air space of 8 cm (3 inch) or more. They will probably do best in soft, peaty water kept at 23-30°C (75-86°F).

Hatchet fishes are a schooling species and appear to feed on insects at the water surface. It is unknown whether they are flight primarily to escape predators or to capture food.

Family – Belontiidae
Order – Perciformes

***Belta splendens* (Siamese fighting fish)**

Size: 6 cm (2.5 inch)

Range: Indo-China and Malaysia

Feeding:

In nature: A wide variety of small freshwater organisms.
In aquarium: Usual aquarium foods including prepared or live foods.

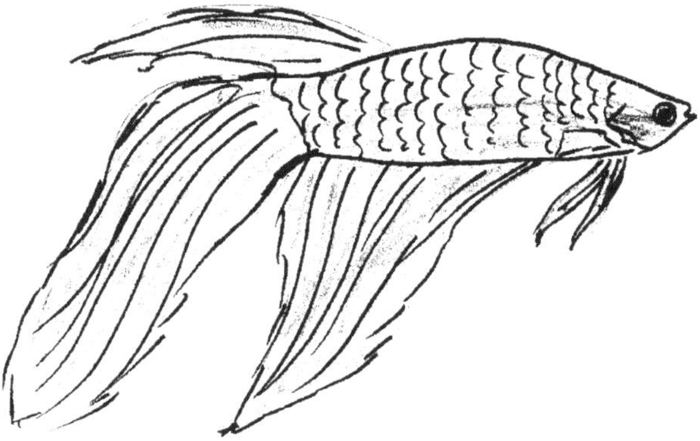

***Belta splendens* (Siamese fighting fish)**

Aquarium care: A single male with one or two females can be kept in a community tank or single-species tank of 40 litres (10 gal) or more. Water quality is not critical but breeding appears to be stimulating by keeping the water level low; such as 15 cm (6 inch) or less. Provide floating vegetation and keep water temperature at 26-30°C (79-86°F).

Family – Pantodontidae
Order – Osteoglossiformes

This family contains only a single species, the African butterfly fish *Pantedon buchholzi*, which is characterized by usual pelvic fins and long fellers and large pectoral fin. The latter resemble butterfly wings, which the fish use to jump out of water, but not for gliding. During the day, butterfly fish usually hide under leaves and floating debris and feed actively only in the evening and at night. They feed upon living animals that live at the surface including terrestrial insects which accidentally land on the water.

Pantodon buccholzi (African butterfly fish)

Size: To 10 cm (4 inch)

Range: Still water and pools associated with rivers in tropical West Africa.

Feeding:

In nature: A wide variety of surface dwelling organisms, including fishes, insects and invertebrates.

In aquarium: A wide variety of food including mealworms, maggots and prepared food.

Aquarium care: Small specimen have been kept in aquaria as small as 75 litres (20 gal), but the aquarium must be well covered as the fish is a notorious jumper. Large aquaria of 300 litres (80 gal) are preferred. This is a peaceful species. Since it lives at the surface, it makes a nice addition to many community tanks.

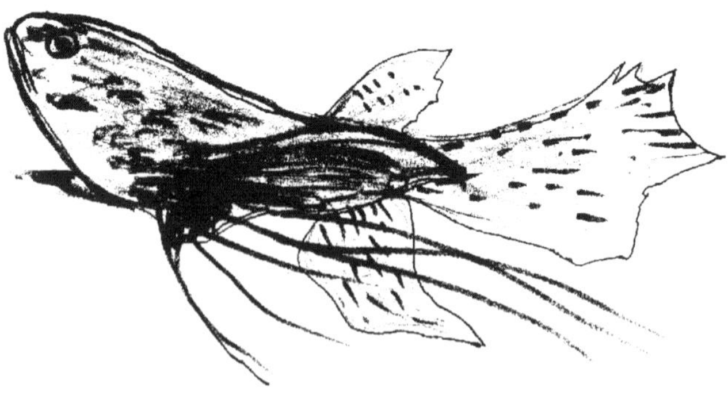

Pantodon buccholzi **(African butterfly fish)**

Family – Mormyridae
Order – Osteoglossiformes

Gnathonemus petersi (Congo elephant fish)

Size: 23 cm (9 inch)

Range: Tropical Africa including the Congo, Cameroon and Zaire.

Feeding:

In nature: Small mud-dwelling organisms

In aquarium: Feed small worms, insect, larval and plankton (*Daphnia*, *Cyclops*, etc.)

***Gnathonemus petersi* (Cango elephant fish)**

Aquarium care: Water quality is not critical, keep between 20°C and 28°C (68-82°F) in a dimly lit aquarium with surface plants, rooted plants and hiding places amongst rocks and wood.

This is probably the best known species of mormyrid to the hobbyist. It does well if kept in a dimly lit aquarium but care must be taken to see that it receives plenty of live food. The Congo elephant fish tolerates other species well and may be kept in community tanks. However, it may react adversely to other species of electric fishes.

Important Marine Fishes

Introduction

Many aquarists like to adorn their tanks with marine fishes which are found in general habitat of coral reefs, coastal sea area and deeper oceanic parts.

The beauty of marine coral fishes commonly kept by aquarists is truly magnificent. The marine fish is new and exciting pass time where few really knowledgeable people exist. The distribution is restricted due to certain basic requirements in salinity, temperature, food etc. Marine fishes are further restricted to certain depths of oceans. Some fishes are adopted to surface water. Evolution has equipped others with to medium depths where a further groups bathypelagic fishes with blacker red colouration, small eyes, wide mouth line at greater depth and soon. By far greatest majority fishes line in on close to continental shelf and the number decreases as the depth increases.

Indeed a great deal of work has to be done before even a basic understanding can confidently be claimed. Like an biological systems nothing in aquarium is static. It is in a static of dynamism with constant change and adaptation. It is upto marine aquarists to offer something for the pleasure he/she receives. Ideally this would be to breed his own supplies of fish and this must be the goal to which all should direct their efforts. Unlike this is achieved the fish in captivity should be treated with utmost respect and the greatest possible care should be taken to prolong their lives or otherwise it may create a problem for conservationalists.

A word of advice for the beginners and aquarists using a new aquarium for the first time, stock of aquarium with a couple of very small hardy and preferably inexpensive fishes in tank; then it should be left for sometime to let it mature before expensive and sensitives fishes are kept.

Classification of Marine Fishes

The marine fishes which are used for small and big aquariums are classified by the taxonomists are as

Order	*Name of Fish*
Heterodontiformes	ex. Horn sharks
Laminifomies	Requiem sharks
Anguilliformes	Moray eells, conger eel, Garden eels
Silurifomies	Catfish eels
Batrachoideformes	Toad fish
Lophiformes	Frog fish Bat fish
Bercyformes	Flash light fish Pinecone fish Squirrel fish

Contd.......

Order	Name of Fish
Syngrathiformes	Shrimp fishes
	Pipe fishes
	Sea horses
Scorpaeniformes	Rock fishes
	Scorpion fishes
	Sceulpins
Pleuronectiformes	Right eye flounder
	Left eye flounder
Tetradonliformes	Trigger fishes
	Pufflers
	Box fishes
	Cow fishes
	Trunk fishes
	Porcupine fishes (spiny puflers)
Perciformes	Sea basses and groupers
	Soap fishes
	Dotty backs
	Tiger perches
	Aholehole
	Cardinal fishes
	Grunts and sweetlips
	Jack and pompanos
	Snappers
	Big eyes
	Croakers and drums
	Fairy basslets
	Spade fishes
	Goat fishes
	Butterfly fishes
	Angel fishes
	Damsel fishes
	Hawk fishes
	Thread fines
	Wrassess
	Jaw fishes
	Parrot fishes
	Thread fin blennics
	Pike blennies
	Clinids

Contd.......

Order	Name of Fish
Perciformes	Blennieo
	Conibtooth bureos
	Pricker backs
	Gunnels
	Gravel divers
	Gobies
	Sturgeon fishes
	Rabbit fishes
	Dragenet
	Mandarin fishes

Some popular Aquarium Fishes

Family – Acanthuridae

Sphaerrua neratopterus

Popular name: Coninion pyjana or Cardinal fish
Distribution: Indo-pacific
Size: 3 inches

Sphaerrua neratopterus

This species of cardinal fish is easy to keep and will soon leave to settle down and enjoy living in aquarium.

It is slow moving fish and although an occasional specimen will accept dried fish food but they prefer some form of fresh protein such as shrimp flesh or beef heart.

The eyes are red and body is quite deep in relation to the length. It is hardy in nature and can be easily included in the list of fishes recommended for beginners.

Sphaeriza oblicularis

Popular name: Motted cardinal

Distribution: Indo-Pacific

Size: 4 inch

Sphaerina oblicularis

This specimen is generally confused with the above specimen, *S. nematoperus*. The colour is more mundane. It is not popular fish for aquariums because it lack colour and becomes active at night only.

It prefers fresh protein food form to dried on freezed dried preparations.

This fish can be recommended for beginners due to its hardiness.

Family – Canthigasteridae

Canthigaster bennetti

Popular name: Bennetts sharp nose puffer

Distribution: Tropical and Indo-Pacific

Size: 2 inches

Canthigaster bennetti

This species is most colourful and suitable for marine aquarium and is very pleasant to look. It is found in relative abundance throughout the Indo-Pacific. In an aquarium it proves quite hardy accepting readily the new conditions and the food offered.

It is peaceful fish like all other members of this particular genera. Requires plenty of space in well aerated tank.

Canthigaster valentini

Popular name: Sharp nose puffer

Size: 2 inches

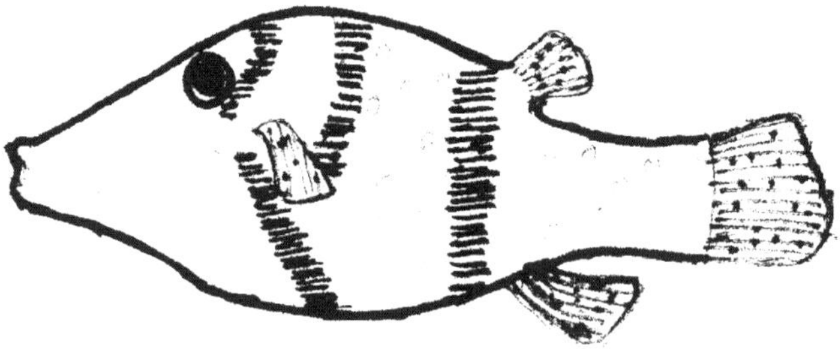

Canthigaster valentini

Body brown in colour with white spots all over. This species is most colourful and suitable for marine aquariums. The fish is relatively scarce and is a excellent community fish providing that fishes it is living are not too large as troublesome.

It make grouting noises if provoked or upset. This popularly called sharp nose puffer and never goes bigger in size, rarely found more than 2 inches in size.

Family – Chaetedontidae

Centropyge argi

Popular name: Pygrey Angel fish or Cherubfish

Size: 2¼ inches

Distribution: Western Atlantic

Centropyge argi

Body very light, blue in colour. This species is hardy in nature. It is quite shy fish, therefore, aquarium should be provided with lot of hiding places. Generally it should be given 3 to 4 days to settle down in its new home after which most frozen or fresh shell fish will be accepted. They can be given in form of chopped mussels, shrimp fish, shell fish, cockles, crab/prawn meat etc. and it will be good if food is given in varied forms.

Centropyge flavissiams

Popular name: Lemonpeelangel or lemon peel

Size: 4½ inches

Distribution: Central Pacific

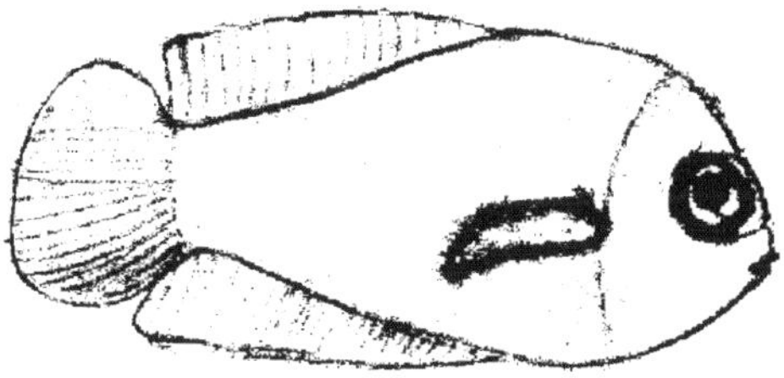

Centropyge flavissiams

A mild mannered fish which will only occasionally scarp with its own kind. Not meant for beginners, some experience is required to keep its successfully. Lemonpeel requires regular feeding of freeze dried or frozen tubfix worms of fairy shrimps. Live brine shrimp can also be used.

The water condition required are pH 8.3, nitrates 0 and temperature 26.5°C is ideal.

Chaetodon citrinellus

Popular name: Speckled Butterfly

Size: 4½ inches

Distribution: Indo-Pacific

Body is brown in colour with black spots and blue coloured band running over the eye vertically.

Although mundane in colouration, is a firm favourite with British Marine aquarists. It is a delicate cooking fish but is extremely

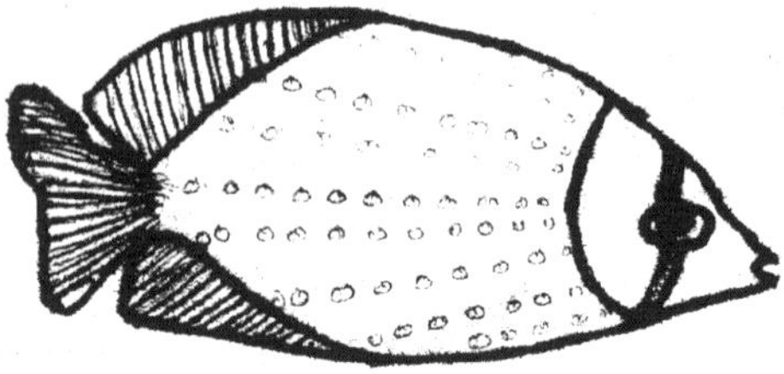

Chaetodon citrinellus

hardy. Specimen of around 1½ to 2″ are normally available and are quite suitable to community set up.

After it's initial shyness have overcome, the fish will flourish. It will readily accept most forms of food, once it has acclimitized itself to aquarium life. One of the easiest butterfly fish to keep.

Chaetodon klieni

Popular name: Sunburst Butterfly fish

Size: 4½ inches

Distribution: Indo-Pacific

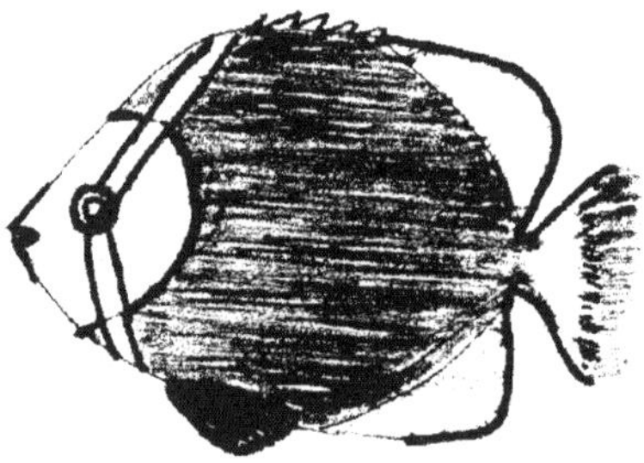

Chaetodon klieni

The mouth and head is white in colour. Body golden in colour with silver spots.

This species is rarely obtainable, despite its abundance in the Indo-Pacific. Food is no problem and all normal aquarium foods are suitable. It is quite hardy and relatively easy to keep.

A black bar runs down from the tip of head down each side and through the eyes, as it does in all members of this genus. It is a small fish but, nevertheless, are worth keeping if the opportunity presents itself.

Adioryx tiere

Popular name: Palau Squirrel fish

Size: 4-6 inches

Distribution: Tropical-Pacific

Adioryx tiere

It is predominantly red with ventral, anal and anterior dorsal fin tipped with white. In certain light conditions horizontal bar may be seen running along the sides of the fish. It is interesting to note here that most nocturnal fishes are red in colour. In aquarium it should have tank mates large enough to avoid being eaten by this predator. Feeding is no problem and can feed easily on fresh form of protein.

Family – Lutjanidae and Serranidae

Mirolabrichthys tuka

Popular name: Purple green / Butterfly perch.

Size: 5 inches

Distribution: Indo-Pacific

Mirolabrichthys tuka

Body very light, green in colour. This species does not grow large. It is exceedingly attractive and easily identified by the probosis like extension of the upper lip. In aquarium it likes diet of canned shrimp, which has been previously washed free of preservatives and chopped. This is a hemaphrodite species. In natural environment feeds on plankton and river it strays from the sheltering place.

Hippeampus trimaculatus (Sea Horse)

The sea horse is a well aquainted fish indeed. Having a head like that of a mythological horse, the prehensile tail can be curled

Hippeampus trimaculatus **(Sea Horse)**

around sea weeds when it wishes to rest. The female passes her eggs (and her responsibility to the pouch in thermal and it is he who, with considerale labour pains, gives birth to the young. A true example of women's lib in the animal kingdom.

Platosus anguillaris (Catfish Eels)

Catfish eels are elongated. Somewhat eel-like fishes with barbels around their mouth. They have two dorsal fins: the first one, along with the anal fin, is confluent with a pointed tail. Sharp serrated

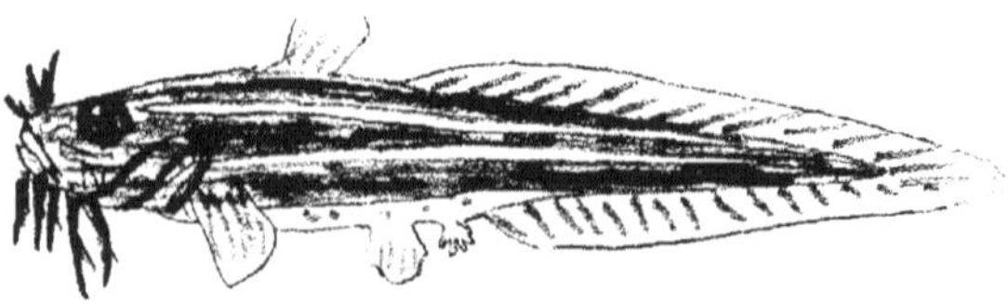

Platosus anguillaris **(Catfish Eels)**

spines, capable of causing severe injury, precede the first dorsal fin and the pectoral fins. Scales are absent. Catfish eels are bottom-living species usually associated with soft substances and often occurring in aggregation. About 30 species are known from the marine and brackish water of tropical India and Pacific oceans and adjoining fresh waters.

Opanus tau (Toad fish)

Toad fish are robust, bottom dwelling fish with large head, wide mouth, strong teeth and dorsally placed eyes. Scales are usually lacking but the body is well coated with mucous. One are more lateral line are present. The pelvic fin inserted anterior to the pectoral fins. The swim bladder is out and is used to produce sound. They are found in coastal bottom-living fishes are found in temperate and tropical water, of the Atlantic, Pacific and Indian ocean.

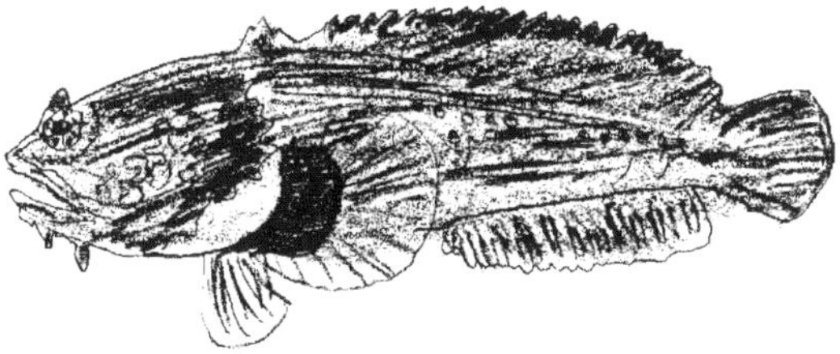

Opanus tau **(Toad fish)**

Lutjanus sebae (Snappers)

They are moderate size fishes. They are more or less compressed and vary from deep bodied to elongate. The dorsal fin is continuous and has strong spines. The caudal fin forked to emerginate. Most snappers are bottom living, nocturnal or crepuscular predator, that prey on fishes and crustaceans. They occur worldwide in warmer seas.

Lutjanus sebae **(Snappers)**

Juvenile of this species are popular with home aquarists. They are good community tank species when kept with fishes larger than bite-size. Adult pale and lose marketing with age.

Pareqes viola (Drums)

They are more or less elongate, compressed fishes, usually silvery in colour as adult. A few species are fusiform in shape. They adopt as bottom feeding habit.

Pareqes viola **(Drums)**

The stripped juveniles appear in schools during the summer. Adults are very secretive during the day and rarely emerge from crevices and ledges, except at night when they feed.

Heterodontus franeisei (Horn Sharks)

Horn shark are relatively small and inactive and usually adapt well to captivity. They have a stocky body with a squarish, blunt; a

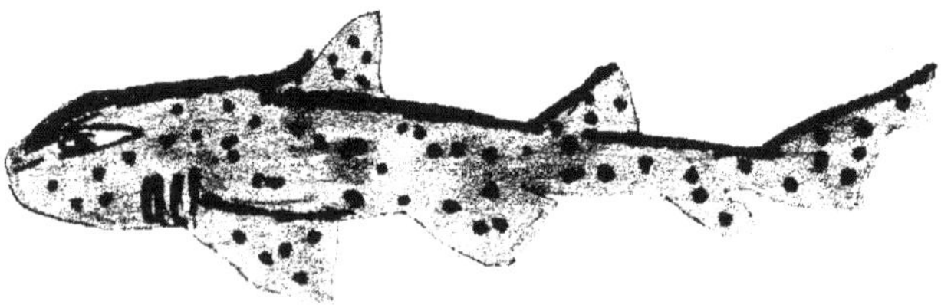

Heterodontus franeisei **(Horn Sharks)**

sub-terminal mouth with small teeth; and two dorsal fins each preceded by a stout spine. The horn shark, are oviparous with internal fertilization, the egg is protected by horny case which differs with each species.

Triakas semifaseiata **(Requiem shark)**

Triakas semifaseiata (Requiem shark)

This is a large family of shark, of various sizes with most species found in the tropics; where they may be abundant. The family includes most species known to be dangerous to human. Requiem shark are fusiform in shape, with a pointed profile. Two dorsal fins, and a caudal fin with an enlarged upper lobe. The spiracles are small or absent. Most species are nomadic. Those that have been maintained in captivity are usually active. Fertilization is internal; most species are viviparous. About 150 species are known. At least one species; the bull shark *Carcharhinus leucas*, ventures into fresh water.

Fish Name, Length and Breeding Temperature [O"con'ell's R.F. (1977)]

Specific name	*Common name*	*Family*	*°F aver. temp.*	*Breed-ing temp.*	*Average l'gth in ins.*	*Natives*
Acanthophalmus semicinctus	Malayan Loach	Cobitidae	78°	-	4	Malay Pen. & Archipelago
Acanthophalmus kuhlii kuhlii		Cobitidae	78°	-	4	Java
Acanthophalmus kuhlii sumatranus		Copitidae	78º	-	4	Sumatra
Aequidens latifrons	Blue Acara	Cichlidae	76º	78º	6	Panama and Colombia
Aequidens maronii	Keyhole Cichlid	Cichlidae	75º	80º	4	Brit. Guiana & Venezuela
Aequidens portalegrensis	Brown Acara	Cichlidae	74º	78º	6	S.E. Brazil
Ambassis lala	Indian Glassfish	Centropomidae	70º	76°	2½	India
Anabas testudineus	Climbing Perch	Anabantidae	76°	78°	10	India, Burma, Siam, Ceylon, Phillipines
Anostomus anostomus	Red Headstander	Characidae	78°	-	4	Guiana
Aphyocharax rubripinnis	Bloodfin	Characidae	72°	78º	1¾	Argentina
Aphyosemion australe	Lyretail	Cyprinodontidae	70°	72°	2½	West Africa
Aphyosemion bivattatum	Banded Fundulus	Cyprinodontidae	70°	72°	2½	Tropical W. Africa
Aphyosemion coeruleum	Blue Gularis	Cyprinodontidae	70°	70º	4½	West Africa
Aphyosemion gardneri		Cyprinodontidae	70°	73°	2½	Tropical West Africa
Aphyosemion gulare	Yellow Gularis	Cyprinodontidae	70°	73°	21	West Africa
Apistogramme ramirezi	Butterfly Dwarf Cichlid	Cichlidae	73º	83°	2	Western Venezuela

Contd.......

Table–contd........

Specific name	Common name	Family	°F aver. temp.	Breed-ing temp.	Average l'gth in ins.	Natives
Aplocheilus lineatus	Panchax	Cyprinodontidae	76°	78º	4	India and Ceylon
Arnoldichthys spiloptems	Striped Panchax	Characidae	78º	-	2¾	West Africa
Astyanax mexicanus	Mexican Astyanax	Characidae	73º	75º	3	Mexico
Badis badis		Nandidae	73°	80º	2 to 3	India
Barbus conchonius	Rosy Barb	Cyprinidae	70º	72°	3½	India
Barbus cumingii	Cumings Barb	Cyprinidae	74°	78°	2	Ceylon
Barbus everetti	Clown Barb	Cyprinidae	75°	84°	4 to 5	Borneo and Malay Pen
Barbus fasciatus	Striped Barb	Cyprinidae	74°	78°	4	Borneo, Malaysia
Barbus gelius	Golden Dwarf Barb	Cyprinidae	70º	72°	1¾	India
Barbus hexazona	Tiger Barb	Cyprinidae	74°	78°	2¼	Malaysia and S.E. Asia
Barbus lateristrigo	Spanner Barb	Cyprinidae	75º	80º	4¾	Malay and E. Indies
Barbus nigrofasciatus	Nigger Barb	Cyprinidae	70°	80°	2	Ceylon
Barbus oligolepis	Checker Barb	Cyprinidae	71°	78°	2	Sumatra
Barbus phutunio	Dwarf Barb	Cyprinidae	71°	78°	2	India
Barbus schuberti		Cyprinidae	71°	75°	2½	South China
Barbus semifasciolatus	Half Banded Barb	Cyprinidae	71°	75°	2½	South China
Barbus setivinensis	Algerian Barb	Cyprinidat	71°	75°	12	North Africa

Contd........

Table–contd........

Specific name	*Common name*	*Family*	*°F aver. temp.*	*Breed-ing temp.*	*Average l'gth in ins.*	*Natives*
Barbus terio	One spot Barb	Cyprinidae	71°	75°	2¾	Bengal, Punjab and Orissa
Barbus tetrazona	Tiger Barb	Cyprinidae	73°	80º	2	Malay Pen. and Borneo
Barbus ticto	Two Spot Barb	Cyprinidae	74°	76°	3½	India
Barbus titteya	Cherry Barb	Cyprinidae	75°	80º	2	Ceylon
Barbus vittatus		Cyprinidae	75°	80º	2¼	Ceylon
Belontia signata	Comb Tail	Anabantidae	75°	78°	4¾	Ceylon
Betta splendens	Siamese fighter	Anabantidae	75°	82°	2¾	Singapore, Malaya, and Indo-China
Botia macracantho	Clown Loach	Cobitidae	75°	-	6	Sumatra and Borneo
Brachydanio devario		Cyprinidae	73°	75°	3½	India
Brachydanio nigrofasciatus	Spotted Danio	Cyprinidae	73°	75°	1¾	N. Indo-China
Brachydanio abolineatus	Pearl Danio	Cyprinidae	71°	75°	2¼	Burma
Brachydanio rerio	Zebra Danio	Cyprinidae	68°	75°	1¾	Bengal
Brachygobius xanthozonus	Bumble Bee	Gobiidae	78°	86°	1½	India and Indo-China
Bryconalestes longipinnis	Long Finned Characin	Gharacidae	75º	78°	4½	West Africa
Carnegiella strigata	Marble Hatchet	Characidae	77°	85°	1¾	Guiana, Central and Lower Amazon

Contd........

Table–contd........

Specific name	*Common name*	*Family*	*°F aver. temp.*	*Breed-ing temp.*	*Average l'gth in ins.*	*Natives*
Cheirodon axelrodi	Cardinal Tetra	Characidae	76°	-	1¾	Upper Amazon
Cichlasoma biocellatum	Jack Dempsey	Cichlidae	74°	77°	7	South America
Cichlasoma coryphaenoides	Chocolate Cichlid	Cichlidae	74°	77°	7	Brazil and Argentina
Cichlasoma facetum	Chanchito	Cichlidae	74º	77º	7	Argentina
Cichlasoma festivum	fcstive Cichild	Cichlidae	74°	77°	6	Amazon Basin
Cichlasoma meeki	Firemoulh	Cichlidae	74º	77°	4	Guatemala and S. Mexico
Cichlasoma nigrofasciatum	Zebra Cichlid	Cichlidae	74°	77°	4	Central America
Cichlasoma serverum	Striped Cichlid	Cichlidae	75°	80º	7	Guiana, Amazon Basin
Colisa chuna	Honey Dwarf Gourami	Anabantidae	78°	84°	2½	India
Colisa fasciata	Giant Gourami	Anabantidae	78º	84°	4½	India to Burma
Colisa labiosa	Thick Lipped Gourami	Anabantidae	77°	84°	3¼	Burma
Colisa lalia	Dwarf Gourami	Anabantidae	78°	86°	2 l	N. India
Copeina arnoldi	Splashing Tetra	Characidae	78°	78°	3	Brazil
Corydoras aeneus	Bronze Catfish	Callichthyidae	75°	78°	2¾	Trinidad
Corydoras agassizi		Callichthyidae	75°	-	3½	Amazon and W. Brazil
Corydoras arcuatus	Streamlined Cory	Callichthyidae	75°	-	2½	Upper Amazon
Corydoras hastatus	Dwarf Catfish	Callichthyidae	76°	80°	1¾	Amazon Basin

Contd........

Table–contd........

Specific name	*Common name*	*Family*	*°F aver. temp.*	*Breed-ing temp.*	*Average l'gth in ins.*	*Natives*
Corydoras julii	Leopard Catfish	Callichthyidae	78°	-	2½	E. and N.E. Brazil
Corydoras nattereri	Blue Catfish	Callichthydae	76°	-	2½	Cent. and N. Brazil
Corydoras paleatus	Peppered Catfish	Callichthyidae	76°	70º	2¾	S.E. Brazil, Argentina
Corynopoma riisei	Swordtail Characin	Characidae	77°	80°	2	Trinidad, Northern S. America
Cryptopterus bicirrhus	Glass Catfish	Schilbeidae	77°	-	2 to 3	Malaya
Ctenopoma acutirostre	Leopard Bush fish	Anabantidae	80º	-	4	Africa
Cynolebias bellottii	Argentine Pearl Fish	Cyprinodontidae	68°	68°	2¾	La Plata
Danio malabaricus	Giant Danio	Cyprinidae	73°	80º	4	India and Ceylon
Elassoma evergladei	Pygmy Sunfish	Centrarchidae	66°	64°	1¼	Florida
Epalzeorhynchus kallopterus	Trunk Barb	Cyprinidae	78°	-	4	Borneo, Java and Sumatra
Epiplatys chaperi	Fire Mouth Panchax	Cyprinodontidae	71°	77°	2¼	West Africa
Etroplus maculatus	Orange Chromide	Cichlidae	74º	84º	3	India and Ceylon
Etroplus suratensis	Striped Chromide	Cichlidae	72º	-	16	India, Ceylon
Gasteropelecus levis		Gasteropelecidae	75°	-	2¼	Amazon Basin and Guiana
Gynanocorymbus ternetzi	Black Tetra	Characidae	76°	83°	3	Rio Paraguay
Haplochromis multicolor	Egyptian Mouthbreeder	Cichlidae	71º	77°	6	Egypt

Contd........

Table–contd........

Specific name	Common name	Family	°F aver. temp.	Breed-ing temp.	Average l'gth in ins.	Natives
Hasemania melanura	Copper Characin	Characidae	72°	78°	2	S.E. Brazil
Helostoma temmineki	Kissing Gourami	Anabantidae	78º	78º	6	C. India and Indonesia
Hemichromis bimacuiatus	Jewel Cichlid	Cichlidae	73º	77º	5 plus	Tropical Africa
Hemigrammus caudovittatus	Buenos Aires Tetra	Characidae	71°	75°	4	Buenos Aires
Hemigrammus erythrozonus	Glowlight Tetra	Characidae	76°	77°	1¼	N. E. South America
Hemigrammus ocellifer	Head and Tail Light	Characidae	76°	78°	1¾	Guiana and Amazon
Hemigrammus pulcher	Pretty Tetra	Characidae	75°	80º	2	Upper Amazon
Hemigrammus rhodostomus	Red Nose	Characidae	75°	-	2	Lower Amazon
Hemigrammus ulreyi	German Flag Tetra	Characidae	75°	-	1½	Paraguay River
Heterandria formosa	Mosquito Fish	Poeciliidae	71°	78°	1	S.E. of U.S.A.
Heteropneustes fossilis		Clariidae	71º	-	26	India, Ceylon
Hyphessobrycon callistus	Jewel Tetra	Characidae	76º	80º	1¾	Brazil
Hyphessobrycon flammus	Tet from Rio	Characidae	70º	73º	1½	Rio de Janeiro
Hyphessobrycon griemi	Red Gold Dot Tetra	Characidae	72°	73°	1½	Brazil
Hyphessobrycon heterorhabdus	Flag Tetra	Characidae	75°	75°	1½	Amazon
Hyphessobrycon innesi	Neon Tetra	Characidae	73°	70º	1¾	Upper Amazon
Hyphessobrycon ornatus	Ornate Tetra	Characidae	77°	80º	1½	Dutch Guiana, Lower Amazon

Contd........

Table–contd........

Specific name	*Common name*	*Family*	*°F aver. temp.*	*Breed-ing temp.*	*Average l'gth in ins.*	*Natives*
Hyphessobrycon pulchripinnis	Lemon Tetra	Characidae	75º	77º	1½	Amazon Basin
Hyphessobrycon scholzei	Black-Line Tetra	Characidae	74°	75°	2	Lower Amazon
Jordanella floridae	Flag Fish	Cyprinodontidae	68°	75°	2¼	Florida
Labeo bicolor	Red Tailed Black Shark	Cyprinidae	76°	-	6	Thailand
Lebistes reticulatus	Guppy	Poeciliidae	74º	78º	1	Trinidad, Guiana, Venezuela
Limia caudofasciata	Blue Poecilia	Poeciliidae	75°	78°	2	Jamaica
Limia melanogaster	Black Bellied Limia	Poeciliidae	75°	78°	2½	Jamaica
Limia nigrofasciata	Hump-back Limia	Poeciliidae	75°	78°	2½	Haiti
Limia ornate	Green Poecilid	Poccilidae	76°	78°	1¾	Haiti
Limia vittata	Striped Mud Fish	Poeciliidae	76°	78°	2½	Cuba
Macropodus chinensis	Round-tail Paradise Fish	Anabantidae	66°	73º	2½	Eastern China and Korea
Macropodus opercularis	Paradise Fish	Anabantidae	68°	75°	3	China
Macropodus opercularis concolor	Black Paradise Fish	Anabantidae	68°	-	3 China	
Megalamphodus megalopterus	Black Phantom Tetra	Characidae	76°	78°	1¼	Brazil
Melanotaenia nigrans	Aust. Rainbow	Atherinidae	75°	78°	5	Australia

Contd.......

Table–contd........

Specific name	*Common name*	*Family*	*°F aver. temp.*	*Breed-ing temp.*	*Average l'gth in ins.*	*Natives*
Mesogonistius chaetodon	Black Banded Sunfish	Centrarchidae	66°	64°	4	U.S.A.
Moenkhausia oligolepis	Glass Tetra	Characidae	74°	78°	5	Brazil and Guiana
Moenkhausia pittieri	Diamond Tetra	Characidae	73º	78°	2½	Venezuela
Moenkhausia sanctae filomenae	Red Eyed Tetra	Characidae	74°	78º	2½	Paraguay
Molliensia latipinna	Sailfin Molly	Poeciliidae	78°	80º	3½	N. Carolina, Florida, N.E. Mexico
Molliensia sphenops	Marbled Molly	Poeciliidae	78º	80º		Texas to Venezuela
Molliensia velifera	Giant Sail-fin Molly	Poeciliidae	78°	80º	4	Yucatan
Monodactylus argenteus		Monodactylidae	80º	-	5	Indian Ocean
Nannacara anomala	Golden Eyed Dwarf Cichlid	Cichlidae	77°	80º	2¾	British Guiana
Nannobrycon eques	Pencil Fish	Hemiodontidae	74°	82°	1¾	Amazon
Nannostomus aripirangensis	Brown Pencil Fish	Hemiodontidae	75°	86°	1¾	Amazon
Nannostomus marginatus	Dwarf Pencil Fish	Hemiodontidae	75°	80º	1¾	Amazon and The Guianas
Nematobrycon palmeri	Emperor Tetra	Characidae	78°	80°	2½	Colombia
Nothobranchius guentheri		Cyprinodontidae	76°	76°	3½	East Africa and Zanzibar
Ococinclus offinis	Dwarf Sucking Catfish	Loricariidae	72º	-	1¾	Rio de Janeiro

Contd........

Table–contd........

Specific name	*Common name*	*Family*	*°F aver. temp.*	*Breed-ing temp.*	*Average l'gth in ins.*	*Natives*
Pachypanchax playfairi	Playfairs Panchax	Cyprinodontidae	75°	75º	3½ to 4	Zanzibar, Seychelles, E. Africa
Phallichthys amates	Merry Widow	Poeciliidae	74º	78º	1¼	Guatemala
Phallichthys pittieri		Poecillidae	74º	78°	1½	Costa Rica, Panama
Phenacogrammus interruptus	Congo Tetra	Characidae	74º	76°	3	Congo
Pristella riddlei	Riddles Pristella	Characidae	74°	78°	1¾	N.E. South America
Pterophyllum scalare	Angel Fish	Cichlidae	78°	80º	5	Amazon and Guiana
Rasbora daniconius	Slender Rasbora	Cyprinidae	74°	78°	3	India and Ceylon
Rasbora heteromorpha	Harlequin	Cyprinidae	76°	79°	1¾	Malay Pen, Borneo, Sumatra, Java
Rasbora maculata	Dwarf Rasbora	Cyprinidae	76º	80º	1	Malaya and Indo-China
Rasbora pauciperforata	Red Lined Rasbora	Cyprinidae	73°	-	1¾	E. Sumatra
Rasbora trilineata	Scissor Tail	Cyprinidae	70º	80º	3	Sumatra and Borneo
Scatophagus argus	Spotted Scat	Scatophagidae	74°	-	4	East Indies
Selenotoca papuensis		Scatophagidae	75°	-	4	New Guinea
Serrasalmus rhombeus	Spotted Piranah	Characidae	72°	-	14	Amazon
Synodontis nigriventris	Upside Down Catfish	Mochokidae	77°	-	2½	Congo

Contd........

Table–contd........

Specific name	*Common name*	*Family*	*°F aver. temp.*	*Breed-ing temp.*	*Average l'gth in ins.*	*Natives*
Symphysodon discus	Discus	Cichlidae	78°	84°	5½	Central Amazon
Tanichthys albonubes	White Cloud Mountain Minnow	Cyprinidae	66°	72°	1¾	Canton
Telmatherina ladigesi	Celebes Rainbow Fish	Atherinidae	77°	80°	2	Celebes Islands
Tetraodon fluviatilis	Green Puffer	Tetraodontidae	75°	-	7	India, Burma, Malay Pen
Thayeria obliqua	Penguin	Characidae	76°	82°	2¾	Amazon
Trichogaster leeri	Pearl Gourami	Anabantidae	80°	82°	4	Siam, Malay Pen, Sumatra
Trichogaster pectoralis	Snakeskin Gourami	Anabantidae	77°	80°	6	Indo-China to Malaysia
Trichogaster trichopterus	Three Spot Gourami	Anabantidae	77°	80°	5	Malay, India, Indo-China
Trichopsis vittatus	Croaking Gourami	Anabantidae	74°	86°	2½	Malay Region
Xiphophorus hellerii	Swordtail	Poeciliidae	72°	78°	3¼	E. Mexico
Xiphophorus maculatus	Platy	Poecillidae	74°	78°	1½	E. Mexico and Guatemala
Xiphophorus variatus	Variegated Platy	Poecillidae	74°	78°	3	Mexico

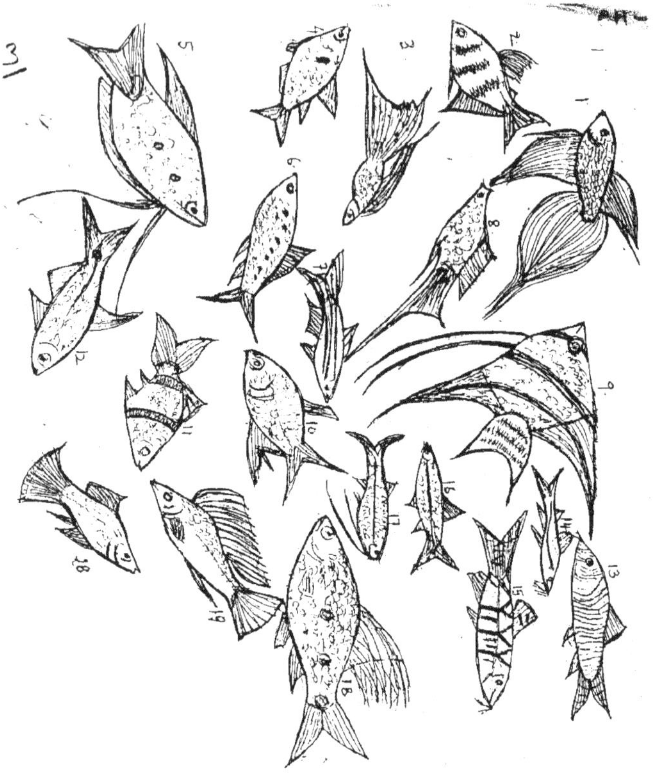

Fig. 23: Ornamental Fishes

1. Siamase fighter, 2. Tetra, 3. Guppy, 4. Serpae, 5. Gourami, 6. Golder Barb, 7. Zebra fish, 8. Sword tail, 9. Angle fish, 10. Lemon tetra, 11. Tiger barb, 12. Tetra, 13. Striped loach, 14. Glass fish, 15. Y loach, 16. *Rasbora*, 17. *Esomus*, 18. Aurali barb, 19. Mollie, 20. Platy

References

Connell, R. F. O. (1977). The Fresh Water Aquarium. First Arco Edition. Arco Publishing Company Inc., New York, pp. 140-230.

Mathew Landau (1973). Introduction to Aquaculture. John Wiley and Sons Inc, p. 369.

Pollution Ecology of Freshwater Invertibrates (1974). Ed. Hart, Jr. C. W. and Samual, L. H. Academic Press, New York & London, pp. 147, 190, 130.

Reginold Dutta (1976). Beginner's Guide to Tropical Fish. Pelham Books Ltd., London, pp. 125-137.

Chapter 7

Food and Feeding

There is a general public concept that good feeding means overfeeding, sadly this is equally true to many aquarists, particularly beginners. Pet fish leaves food and it falls to the bottom, here to decompose, producing a variety of toxins, which effect readily and adversely on the health and well-being of the fish. The fish become lethargic at first and if overfeeding is continued, pollution may occur so that the captive inmates die.

Aquarium fishes have become adapted to a variety of food. Some of them feed exclusively on plants, other feed on animals while a large number of species are omnivorous deriving their requirement from both animals and plants. A wide range of organisms are used as food by fishes. This also ranges in nutritionally efficient artificial food. Zooplankter includes various kinds of protozoa, micro crustaceans, microscopic invertebrates, larval and eggs of various animals. Among larger organisms consumed by aquarium fishes are annelid tubifex and blood worms, snails, mussels, crustaceans, insects, smaller fishes and even frogs and tadpoles meat. The plant material includes unicellular algae, and soft portion of higher aquatic plants. These fishes also vary in their feeding habit according to age and certain adaptive morphological features. To be safe from these inconvenience, the feed of aquarium fishes and its constituents should be changed from time to time to make them healthy and active is aquarium.

To provide a complete nutritious diet it is necessary to maintain the aquarium well and recently the emphasis has been given on quality of food rather than the quantity. An effective ration for aquarium fishes is that should be nutritionally adequate. It must also be palatable to the particular fish thus the constituents may be in quite good proportion with special reference to size, taste and nutritional status. The food should be made up of the following requirements.

Protein

It is required by fish for proper growth and maintenance. All aquarium fishes require arginine, lysine, histidine thaimine, methionine, valine, leucine, isolusin, phenyalanin and tryptophan. At the same time the proper amount of non essential amino acids are needed. For this meat and bone meals, beef (animal source) and cotton seed, soybean meals (plant origin) and microbial source of protein are also used. Ornamental fishes need 30-40% of protein contents.

Lipid

Lipids are needed for caloric contribution in aquarium fishes and also provide essential component of bio-membrane and cellular organelles. It is excellent energy source of most of the aquarium fishes. Channel cat fish (*Ictalurus punctatus*) has been successfully grown on diet containing upto 10% fat added either from beef, olive oil or *mehaden* oil. High melting point animal fats have recently been shown to be useful energy source for cold water aquarium fishes. For normal growth and good health turbot must be given long chain polyunsaturated fatty acids of w_3 and w_6 series in diet.

Carbohydrate

Many of the aquarium fishes are carnivorous. It has been proved that precooked starch is much more readily utilized by rainbow trout than other trout. Some aquarium fishes has been observed to take chitin. Chitin is digested by some bacteria present in gastrointestinal cavity.

Vitamins

Like other vertebrates these fishes require absolute amount of water soluble and fat soluble vitamins both, so add vitamix (1%) in fish diet.

Minerals

A considerate amount of minerals is needed in diet. The main required minerals are Ca, P, Mg. The above help in increasing growth rate of aquarium fishes. Add mineral mix with pigments in pellets.

Though fishes like to have food, it must be always available to them, in natural state but it is difficult to arrange in aquarium. Thus foods is provided to them time to time in a day and never more food is to be given. Food for aquarium fishes are categorized in two:

(a) Natural fresh or alive food (b) Artificial, prepared or dried.

It is always better to change the type of food provided according to need and environmental condition of the ornamental fishes.

Natural Food

Traditional live food keeps the fish in good health and a well balanced diet. Since they, not only have nutrient value but the fish also enjoy catching them, so gaining psychological benefits as well. The common food include: ciliates, flagellates, rotifers, worms etc.

Ciliates like *Paramecium* food on dissolved nutrients in water and reproduces by binary fission.

Flagellates

Smaller flagellates of microscopic size are *Euglena* and others. These are necessary to obtain good rotifers, mostly used for diet of fry-fish. Only free swimming kind of rotifers are important for aquarium fishes.

Plant forms

Algae zoospores, these are extremely minute zoospores produced by filamentous strand like algae are ideal food for rotifers and *Daphnia*. These spores after two days become filamentous. Filamentous algae are useful for fry and *Daphnia* etc *Volox globator* which rolls about in water like a tiny green bubbles and bright green *Pandorina, Pediastrum*.

Desmids and Diatoms

Desmids are rich source food for insect larvae, *Daphnia*, baby fishes and also oxygenate water. *Gammarus pulex* of 1½ inch is delicious food for larger carnivorous such as *Cichlids*.

Aquarist should care for ideal live food, is easy to handle and can be obtained in good quantities. Recently a good live food *Artemia spp (Artemia salens, Artemia monica)*, brine shrimps for ornamental fishes are innovated and are also assessable throughout the year.

Blood Worms

These worms are resemble with the larvae of *Chironomous* and can be fed to the larger fish without chopping or shredding. These can be obtained easily from the surrounding ponds and is always found with *Daphnia*. One can cultivate them in mud.

Other Kinds of Food for Fishes

Fresh Water Shrimps (*Gammarus pulex*)

Fresh water shrimp are ideal for the larger aquarium fishes. They grow to about ¾ inch and since they feed on decaying animals and vegetable matter, help to clean the tank by scavenging.

Brine Shrimp (*Artemia salina*)

Brine shrimp are one of the most useful and amazing additions to the aquarist pantry. The eggs, which in themselves are not a food, are obtained in a vial and put under reasonable storage conditions. They can remain dormant for many months. When newly hatched, they provide an excellent early live food diet, and can be given to larger fish. For feed, the shrimps can be siphoned off or straining through several thickness of muslin which is then rinsed in the aquarium. The former is the better method because only shrimps are collected from old eggs. Precaution should be taken, never pour the salt solution into aquarium since it upset the pH and hardness of water.

Some liquid proprietary foods contain particles of food that is particularly beneficial to rearing brine shrimps.

Cyclops

These are small crustaceans. They have the appearance of possessing only one central eye. These crustaceans are named *Cyclops* after one eyed giant of mythology .They are usually found in ponds containing *Daphnia* and can be collected. They can be cultivated in a shallow container of *Infusoria* water where they will breed rapidly. These tiny crustaceans can be fed to all aquarium fishes and are particularly useful for small fry just out of the infusoria stage providing they are graded to a suitable size by sieving. *Cyclops* developing double sac containing eggs are situated at the side of body and add to their desirability as a aquarium fish food.

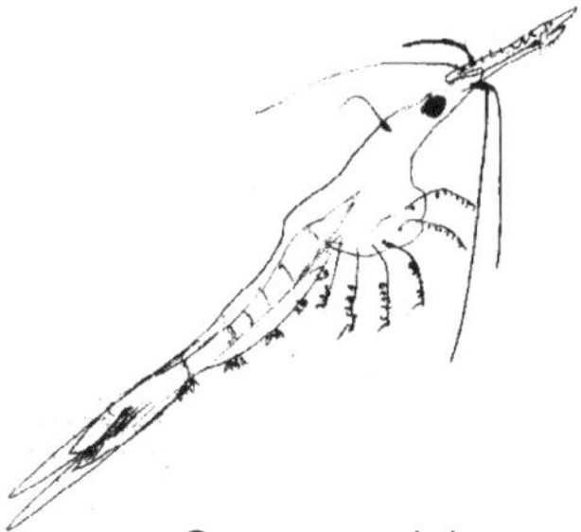

Oppossum shrimp
Prarunus hexuosus

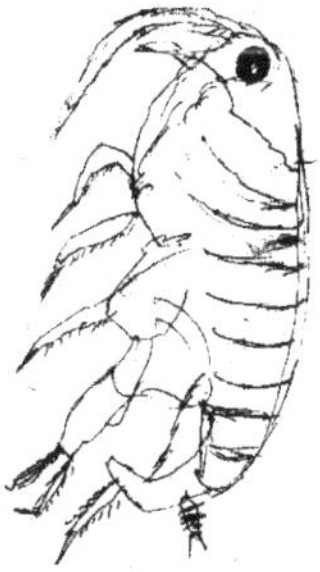

Sand hopper
Talitrus saltator

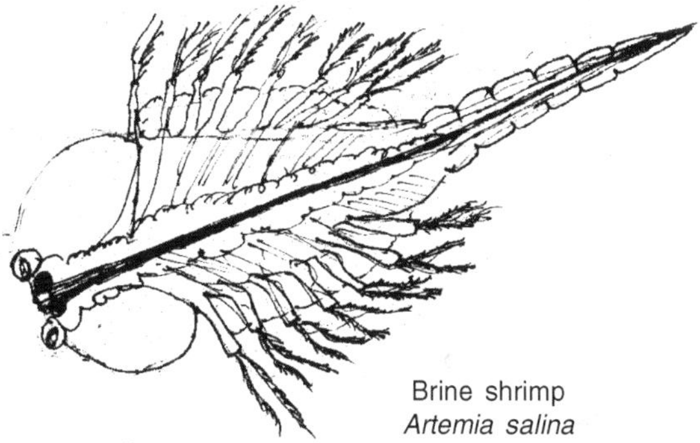

Brine shrimp
Artemia salina

New hatched brine shrimp

All are good seahorse fodder.

Daphnia

It is most universally accepted and best known of all the live foods. They are found in almost every country in dirty ponds, particularly in the ponds where certain manure is present applied. They appear as a cloud of bugs moving. The colour of crustaceans vary from grey, green to red. Latter is more sought after they are thought to be more nutritious. *Daphnia* found abundantly between April and October but also found in winter in isolated batches. It is advisable to only feed small quantities at any one time, so that they are easily consumed and any dead ones remaining should be siphoned off, as they spoil the aquarium water. *Daphnia* can be stored for a few days. They are kept cool, in reasonably large containers, Metallic container is never be used. Aquarist should prefer to use ceramic, porcelain or even old porcelain tank, enamel baths, plastic or glass and use water (tap water) free of chlorine. The best way to transport *Daphnia* is on frames made of wood with an absorbent material sketches across. The material is soaked in water and *Daphnia* spread thinly over the material.

In case, if *Daphnia* is collected by aquarists oneself, it should not be given of fishes directly, but let them stand for a day or two so that undesirable creatures and other dead specimens can be removed. Draw off the healthy *Daphnia* with a siphon tube and strained through a net sieve and simultaneously graded the *Daphnia* for small and large fish to be reared. If supply of *Daphnia* is not in adequate amount the aquarist can culture it easily. Take a container of approximately 14 gallons with a water depth of 6 inches. Prepare each container with a handful of old sheep or cow droppings or chicken manure. Leave it for 14 days then add little baking yeast, and a little animal blood. After a few days add *Daphnia* when the water becomes denuded of nutrients feeding should be renewed. Preserved *Daphnia* are also available in aquarium shops just put them in the water for ten minutes before use.

Earth Worms as Aquarium Fish Food

These worms are favourable food and accepted to most of the aquarium fishes. They are chopped up or shredded before use. An old worm pair cut by scissors into different sizes and given to desired fish or available fish. Earthworms are kept in aquarium as feed only, in case other foods are not available, easily. The worms should be cleaned well before shredding. Aquarist should not use worms that exude a yellowish secretion, the aquarium fish will not eat them and ultimately it will pollute the water. Nottingham red worms and other red worms found in compost heaps are one of the best food.

Gentles

These are excellent food for the larger species of fish and specially favoured by cold water aquarium, fishes. Gentles are the larvae of bluebottle or blow fly pale, whitish grubs about ½ inch long. Due to their larger size they quickly satisfy the appetite. One precaution for aquarists is that coloured gentles are not used, only naturally pale coloured gentles are given as food to fish.

Glass Worms (*Chaoborus* or *Corethera* Larvae)

These are also called ghost worms. They are transparent creatures of about ½ to ¾ inch in length but without the blood of course. These can easily be purchased from suppliers of fishing tackle. They are also found in ponds well shaded by the overhanging foliage. These organisms are tiny and transparent. At the time of netting, net must be swept through the water swiftly and steadily. The glass worms lend themselves quite well to storage in small aquarium. These worms are also easily available during winter.

Infusoria

It is a loose term to describe in the water organisms small enough to be fed to the very tiniest fish. It is a type of 'Primeal soup' containing the necessary ingredients for the first link in life chain. It is prepared from squashed lettuce leaves placed in a quarter of water, which is then left out in sun and air. The minute *infusoria* spores will settle on the water from the atmosphere and multiply on the products of decaying lettuce. Infusoria can be promoted by putting a few infusoria with snails (*Ampullaria paludosa*). Their partly digested droppings provide the essential food. These snails feed on lettuce. Excellent cultures can be made from duckweed and lettuce and other leaves. Dry the leaves thoroughly in an oven and store in clean jars. It can be used whenever required.

Richness of culture can be checked, by placing a spot of water taken from the top surface of the culture, on to the slide under a low power microscope. It should be teaming with minute life about the fine dust size.

Micro Worms (*Anguillula silusiae*)

These are ideal baby food and fed to fry just out of the *infusoria* stage. These are tiny almost invisible worms. Culture of these worms can be purchased from local dealer. The culture media is prepared from soyaflour, semolina, or oat meal. Oat meal should be mixed with water, 1½ oz to a half pinch and then cooked and left to cool. The culture is placed in food mass to multiply. Fresh supply of these micro worms is useful for the aquarium fishes otherwise they may deteriorate naturally for better collection of worms. Match sticks are soaked thoroughly in water a few minute before use.

Mosquito Larvae

These larvae are only to feed fishes, over 1.5 mm in length. Smaller fish are liable to choke. Mosquito larvae can be collected easily from ponds puddles, water logs, ditches and almost in any container near dwelling houses. The larvae are wriggling creature about ¼ inch in length and have a knob like head and 'Y' shaped tail. They remain near the surface of water. They can not utilize the dissolved gasses in water, like fish. So they are compelled to utilize atmospheric air. This, they do by means of a tube situated on the last abdominal segment the tip of which they thrust out just above the water surface.

Rotifers as Fish Food in Aquarium

Rotifers are an ideal early food for some of larger species of aquarium fishes and will thrive in an infusoria culture as their main nutrient is unicellular algae. These are minute aquatic animals feeding and swimming by means of bands of beating cilia revolving wheel like hence named animalcules. An aquarist can obtain rotifers in pond usually among weeds. These are collected by extremely fine net, or very fine muslin. One species *Brachionus rubens* is often found adhering to the body of *Daphnia*. These can be removed by washing the *Daphnia* in a fine sieve with strong jet of water. These rotifers are knocked off the pass through the sieve leaving the *Daphnia* for feeding to the larger, aquarium fish.

Tubifex Worms as Fish Food in Aquarium

It is good live food source and can be obtained all over the year. Their length varies from 3/8 inch to 2 inch long. They are extremely thin and rusty red in color and found in plenty of mud, streams and ponds, collected by lifting the mud with a spade and putting it into muslin bags. Then swell the bags in streams remove all the worms. They form a ball like mass in shallow container.

Worms are stored in a bucket or large bowl into which a slow steady streams of water is allowed to enter. If any time worms become bunched jet of water is poured, for separating them.

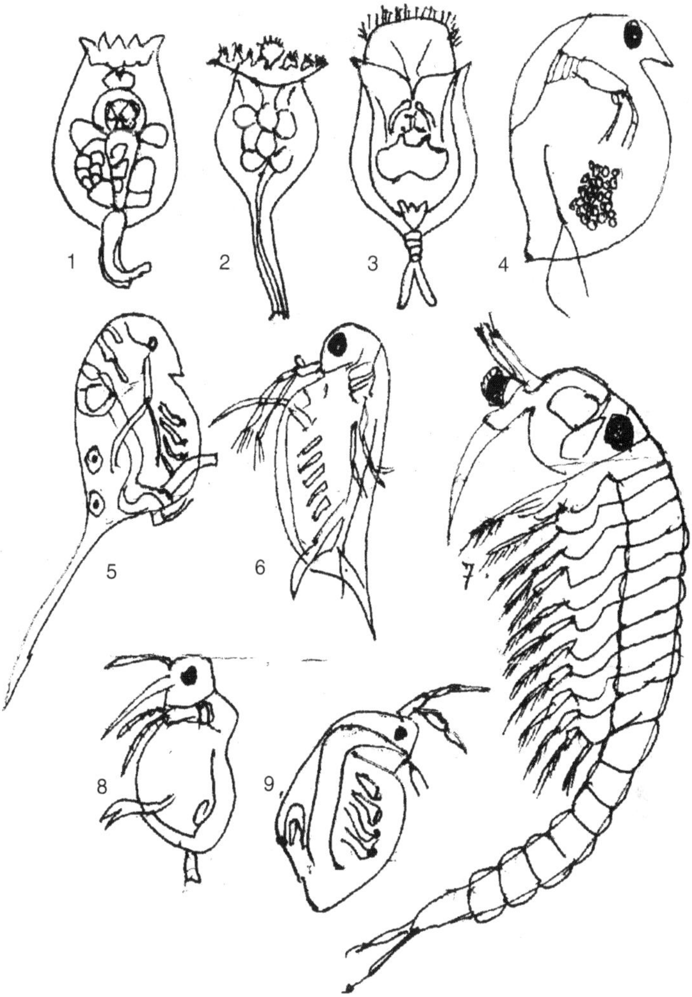

Culcutrable fish food organisms: Rotifers

1. Brachionus plaicalitis; 2. Bachionus rubens; 3. Euchalnis spp. cladocerns; 4. Daphnia carinata; 5. Daphina carinata; 6. Moina spp.; 7. Moina; 8. Ceriodaphnia spp.; 9. Artemia.

A bunch is dropped into aquarium, they will eventually scattered and bury some part of themselves in the gravel, the protruding ends waving around as if in strong breeze. Fish eating these individual swarms usually only manage to bite a section off leaving a part of worms in the gravel to decay and cause problem later. These live food of aquarium fishes are greatly available in market.

White Worms (*Enchytraeids*)

These are soil dwelling creature and very similar to earth worms. White worms cultures are available in winter as supplementary food, they are easy to breed, they provide food that is highly nourishing. The worms obtained from stores (market) are placed in the depression and mesh spread over them, and they multiply at temperature near 15°C and when the soil is wet. After boxes are set up, rotate them for feeding the aquarium fish. They may get culture time to reproduce and ensures an adequate supply. Before feeding, worms should be *free of soil particles* and give quick rinse in water. This will remove any adhering particles or take a few and place them in shallow container such as food container lid and place where it can receive a gentle heat. Worms will amalgamate into a ball for making collection easy.

Dry Food

These are high quality feeding stuffs, contains of many varieties. Flake food can be fed as a staple diet to a large variety of tropical aquarium fishes without any additional food. The fish should be fed 2 to 4 times a day so as much food will be consumed in a few minutes. Flakes are liked by surface feeders, other prefer to catch the flakes as they sink.

Artificial Food

There are a number of recipes for preparing food in by artificial means. Some are given as follows:

(a) Home made

This can made in the kitchen very early by following method:

One cane of cat or dog food, 1 packet of wheat germ food, such as Bemox, 2 teaspoons calcium carbonate, 2 teaspoons honey, 1 packet of pure gelatin crystals and add cheese, roughly equal in bulk to the canned or baked meat, ½ cup approximately of green soft vegetable, such as spinach and 2 eggs. Because aquarium keeping is a luxury and hobby, their food should be of good quality and expensive of course.

Strain any liquid from the canned meat and, it is possible to reduce it to a fine mesh, grind the cheese and simmer in water until the cheese has melted, then add the meat. Stir to prevent the mixture sticking or burning. Remove from heat and stir in two teaspoons of honey, two teaspoons of calcium carbonate and thoroughly blend together. Add some natural pigments, canthxanthin, astaxanthin, sugar beat, chilli 2%, vitamix 1%, mineral mix 0.8%. Add the two eggs and continue to mix whilst slow adding to wheat germ until the mixture becomes very thick and doughy. Finally melt the gelatine crystals into a half cup of hot water, stir until it fully dissolved, beat into the dough thoroughly. The food can than be placed in suitable containers and set aside to cool till it set into a ruspberry texture. It can be stored indefinitely in a deep-freezer.

(b) Gordon's formula

Remove the blood vessel's connective tissue and other fibrous material, 0.45 kg of beef liver, and chop into 13mm cubes. Taking small quantities of liver at a time, add equal amounts of cold water and blend them in a high-speed blender. The resulting liquid should be strained into a large bowl. Add 2 teaspoons of non-iodised salt. Now add 20 teaspoons of pre-cooked baby cereal such as porridge oats, and then stir thoroughly. Fill small glass containers, such as baby food jars, but do not seal them. Place them in a pan of water, to the boil. Allow the jars to remain in the water for half an hour after the heat been turned off. When they have cooled, seal and keep in refrigerated or frozen.

(c) Some other

(i) A liver-based dried food used as the American museum of Natural History and the New York Aquarium. It is composed of 2.3 kg of beef liver, 6.4 kg of wheat germ, 2.7 kg of shrimp shell meal, 2.7 kg of shredded of shrimp meat, 1.4 kg of lettuce and 1.4 kg of spinach. The liver is cut into 2 inch pieces, placed in pan, covered with water and boiled for 15 minutes. Ground finely and put back into the pan with the remaining ingredients, and the entire combination is boiled for an additional 15 minutes. The resulting warm paste is then spread thinly to dry, finely broken and ground to size.

(ii) Use two tumblers of powdered puppy biscuit, one tumbler of powered dry shrimp, and 3 pinches of salt, mix together and moisten with water which has an egg beaten into it. Add enough water and egg to make the mass into a stiff but workable consistency. Spread in pans to dry in quarter inch (6 mm) thick sheets. When dry break into pieces and grade into size.

(iii) Use a quarter tumbler powdered cod, three tumblers wheat flour, three-quarters of a tumbler of powdered shrimp, 3 teaspoons baking powder, one teaspoon of Epsom salt and three teaspoons of powdered chalk, make into the consistency of bread dough with two raw eggs and water. Place in a pan and bake, like bread, in an oven. After baking, allow to cool and cut into thin slices. Allow slices to dry, break into crumbs and soft to size, as desired.

(iv) *Freeze dried foods*: There are various live food which are quickly frozen at sub-zero temperatures and vacuum-removing the crystallized frozen water. This method of preparing the food ensure the cellular freezing techniques. It has been found that freeze, drying of foods preserves the essential nutritive values and the food can be stored, in suitable containers, for an indefinite period of time. Within the range of freeze dried food will be found *Daphnia; Tubifex*, Blood worms and others suitable organisms for feeding to gold fish and other aquarium fishes.

(v) *Fresh frozen foods*: The fish keepers can obtain "Frozen food and meals" for the aquarium fishes, available food include brine shrimp, mussel, brown shrimp, lobster, cockle, *Daphnia*, Blood worms, beef and lobster eggs. These foods are treated with gamma-rays which ensures that they are completely free from any diseases. After this vitamins are added such as A, B_1, B_2, B_6, B_{12}, D_2, D_3, E, F, H and K. They are, therefore excellent foods, being pure protein and wholly edible. If a deep freezer is available the food can be kept safely with the domestic food and supplied until required. It is only necessary to break off a suitable portion and return the remainder to cold-storage.

(vi) *Commercially prepared dried foods*: Nowadays it is possible to obtain a wide range of excellent commercially prepared dried food. Granular feeds are available ready graded to suit various sizes of fish. Pelleted feeds are manufactured especially for feeding pond fish; they have the ability to remain buoyant and float upon the water surface for considerable time. Many pond keepers prefer this form of food because it brings the fish to the upper water where they can be easily seen and enhance the beauty of pond. Possibly the greatest invocation has been in the production of "Flakes foods". Flake -foods are tissue thin flakes, one leading British manufacture states, has especially formulated and manufactured directly from whole fresh ingredients based upon eight different basic formulae. Now available in all over the world and India also. The flakes vary in colour according to the individual formula, pink coloured flakes are based upon white fish; green flakes contain vegetation such as kelp and spinach, light brown based on fresh meat and liver plus blood and ground bone, yellow part of high protein in soya and egg; red combines fish, mixed cereals and blood; organic is described as a colour food for enhancing the colour of fish and contains canthaxanthin; dark brown has meat offals, blood and ground bone; and black has extra vitamins and minerals in its make up. Aquarium fish appear to accept these flake food readily, which float for a short time before slowly sinking through the water. They are, therefore, a useful form of alternative feeding.

Same type of food is also developed with different gradients of GBPUAT Pantnagar-263145, India.

Feeding

The frequency and amount of feeding is dependent upon a number of fishes with special reference to quality and temperature of the water. The gold fish will have little appetite if the water is polluted or the amount of dissolved oxygen is low. It must also be borne in mind that as the temperature falls so the appetite of the gold fish will diminish until, at around 40°F (4.4°C) the fish ceases to feed. Therefore, gauging the amount of and frequency of feeding becomes a matter of practice and experience. However, observation will act as a safety guide in the matter. The sensible approach is often the fish just give a small amount of feed and keep watch to ascertain how quickly they consume it. If the food is eaten greedily with no signs of the fish slowing down or losing interest, add just a little more, continue to do this until the fish show that their appetite is being lost or satisfied. It will not too long before it become possible to gauge the correct amount of food often at any given time. Now a days demand feeding is the concept, fill a container with food and arrange tubes fish will suck the feed whenever needed. Use feeding cup for worm feeding. For dry food use feeding ring and put a bowl/ pot just beneath the ring at bottom so feed if sinks come into the bowl/pot.

There is a danger in overfeeding, not because of the fish will eat too much, but because of the risk of the uneaten food easing the water to be polluted. Of course, certain live foods do not create such problem although a great many *Daphnia* can deplete the dissolved oxygen content of the water, and so cause the fish to suffer. Few fish have died from under feeding, although the rate of growth may well be stunted and the fishes become less resistant to minor disease. During the warmer months of the year, fish are able to digest the food quite rapidly and, to obtain maximum growth. It should be fed three to four times daily. In fact, if a light is left above, then quarters they will feed in the night also, however, intensive feeding (which force the growth) is neither recommended, nor it is necessary.

As the year progresses the water temperature will become lower, and the speed with which the fish digests its food will being slow. The frequency and/or quantity of feeding must be adjusted accordingly. Again the appetite of the fish will indicate the corrections needed to the feeding schedule. Fishes those are kept in fish house tanks or outdoor ponds will exhibit these changes much more positively than those kept in the more protected environment of the indoor aquarium. During the autumn it will be noticed that the appetite of the fish will revive. This is an instinctive desire to put on extra fat in preparation for the winter fast, and should be catered for. During the cold months the aquarium fish will become dormant when the water, cools very much. At these low temperature the digestive system of the fish is unable to assimilate any food, and the fish lives off the accumulated body fats and reserves the attempt to feed at such a time is both wasteful and dangerous. Any food in the digestive tract is likely to decompose and may well result in the death of the fish. It is, therefore, essential that the fish is encouraged to build up a good reserve of body fat to enable it to survive in the hard period. Plenty of good nourishing food, such as earthworm and other fleshy muscles, should become a prominent part of the diet. In this way the fish will be assisted to develop the very necessary reserves of fat. In spring temperature will begin to improve, and as the water slowly warms the fish will again being to show an interest in food. Due to the winter fast the fish will neither be able to digest nor will it require, large meals. Commence feed by offering small amount of *Daphnia* and similar live foods, Gradually increasing the feeds to include prepared foods and live feed improves the appetite of the fish.

The wild fish have period of plentiful feeding and a period when food is very short supply, the wise aquarist will endeavours to follow nature, in feeding arrangements and will recognize that at what time to feed and to withdraw from feeding.

Healthy fish in the pinkish glow condition are more likely to spawn and give a higher % of fertilized eggs than fish in indifferent conditions. All fishes should be kept in ideal conditions but it is worthwhile to treat fish intended for breeding to a period of conditioned feeding for about a week or longer before spawning is attempted.

Feeding Young Fish

Most of the feeds required to rear fry have been mentioned above but for general guidance of the aquarists following table is helpful:

Small sized (Anabantids)

Stage I	*Stage II*	*Stage III*	*Stage IV*	*Stage V*
Green water Flour size Dry food	Infurosia	Infusoria Microworm Rotifers	Brine shrimp, Sieved *Daphnia* Chopped White worms	Larger *Daphnia* small flakes food

Medium sized fry (Characins, barbs)

Stage I	*Stage II*	*Stage III*	*Stage IV*	*Stage V*
Egg yolk (hard boiled & water mix) rest same	Fine dried food, Fine graded shrimps and rotifer	Brine shrimp Micro worms Sieved *Daphnia*	Flake of medium size *Daphnia* Chopped dry worms	Adult *Daphnia* Flake dry food

Large size fry (Cichlids, live bearer)

Stage I	*Stage II*	*Stage III*	*Stage IV*	*Stage V*
Brine shrimps, sieved *Daphnia*, flakes Micro worms, rotifers	Medium *Daphnia* Chopped White worms	Chopped white worms Fine flakes and dry foods	Adult *Daphnia*	Larger dry food and flakes chopped tubifex

When feeding, flour size dry food never sprinkle on water surface, it is most difficult to assess how much to feed. With the result that an excess is usually given, some of which will remain uneaten and cause pollution. So to feed, dip the bag into aquarium the food will disperse into the water in a fine cloud. Apply feeding ring two or three in a big aquarium to make a limited area for fish, this is a method of feeding in control. In another method fish should never be given more than they require, quickly eat in 10 minutes. Fish may be fed 3 times daily one feeding in morning, mid-day and next a few hour before the sunsets. To avoid pollution excess feed is removed by siphons or net. Thus, food and feeding of aquarium fishes is of the basic and primary importance. It is the aquarium, for aesthetic as well as study purposes and to maintain it healthy and unpolluted the good and proper technique should be followed.

Change in daily diet with alterations in pelleted and natural live organisms are much more important.

Menu chart for summer and winters are given below:

Spring and Summer menu

Sunday	a.m. *Daphnia*	p.m. *Tubifex* or earth worm
Monday	a.m. Blood worm	p.m. Home made food
Tuesday	a.m. Earth worm	p.m. Pellet food with algae
Wednesday	a.m. Flake-food	p.m. Blood worms
Thursday	a.m. *Daphnia*	p.m. Porridge oats
Friday	a.m. White worms	p.m. Flake-food + Spirulina
Saturday	a.m. Earth worm	p.m. Home made food

Autumn menu

Sunday	a.m.	Porridge oats
Monday	a.m.	White worms
Tuesday	a.m.	Flake-food + Spirulina
Wednesday	a.m.	Boiled fish
Thursday	a.m.	Home-made food
Friday	a.m.	Finely chopped cooked liver
Saturday	a.m.	*Daphnia* or earth worm

These menu has been based upon the more easily obtained foods and, of course, the cooler temperatures of autumn, menu only one feed each day is required. The novice should use discretion and adoption of the suggested feeding schedules according to the available food supplies and prevailing water temperature.

Live Feed Culture for Fishes

The first rule, to remember about feeding is that, over feeding is one of the most common causes of pollution. Foods with a high protein content are the most satisfactory, as they are more fully used up by fish. Dry foods especially formulated, are richer than the live feeds but spawn, fry and ornamental brood fishes (ready for breeding) prefer live feeds rather than formulated feeds. Further the minerals, vitamins and micronutrient needed by the spawn and early fry are not yet fully understood to be incorporated adequately. The natural live food organisms supply these and thus compensate for them when formulated feeds are used resulting in healthy growth and vigour of fry and brilliant, brightness in fishes. Therefore, it is always preferable to have a regular supply of live feeds. Live food organisms are also serve as living capsules of nutrition. They are composed of protein, carbohydrates, lipids and contain different types of minerals and vitamins.

Culture of Diatoms

It involves a number of stages, except the last stage i.e in the large scale batch culture or mass culture, all the stages are invariably produced successively in air-conditioned rooms. Culture of fresh water diatoms such as Chlorella, Navicula and *Cyclotella* can be undertaken in simple test tubes, to large out door tanks.

Culture in Test Tubes or Petri Dishes

In each of the sterilized test tubes, a known quantity of the medium is introduced aseptically and the tubes are cotton plugged similarly a pair of suitable petri dishes one fitting over the other, is taken and in the lower dish, a known quantity of medium is introduced aseptically and covered immediately with the other petri dish. The live diatom cells are sorted out from plankter sample obtained by filtered pond water with a fine net with the help of binocular microscope and pipette these cells, identified and washed repeatedly in the sterilized nutrient medium, in five to six watch glass. The cell after repeated washing are then transferred to test tubes or petri dishes containing the medium. It would be better, if the test tubes or petri dishes are agitated preferably by mechanical means so as to keep cells in motion further. Culture should be exposed to light from a fluorescent tube. The sub-cultures are made periodically and culture, of test tubes are then transferred to flask (1-3 litre) containing the nutrient rich medium. Before inoculating, the flask with test tube cultures it is necessary to record the number of cell/ml and volume of culture. When cells are in exponential of growth, the culture should be harvested and transferred to larger culture tanks. The flasks after inoculation are shaken daily. They are not agitated by mechanical device. Sub cultures in flasks may be made once in every two weeks. Bacteria free culture of diatoms is not possible owing to practical difficulties. Mono culture of diatoms species with the possible contamination of bacteria is easier and such cultures some are ideal and staple food sources for zooplankter, cultures include larvae, indoors and outdoors. If suitable laboratory conditions are provided, diatom culture could be maintained throughout the year.

Sterilization of Glass Ware

The cotton plugged glass ware of standard glass such as petri dishes (5-10 cm diameter), test tubes bottles flasks (1-3 litre) and carboys (20 litre) are first cleaned with chromic acid, chlorinated sulphuric acid followed by repeated rinsing in distilled water and sterilize in an autoclave for a period of 30 minutes under a pressure of 7 kg/cm^2.

Preparation of Medium

The culture medium is prepared with various salts as shown in table and is sterilized in autoclave, cooled and kept ready. Concentration of various salts required for the medium for diatom culture is given below:

Medium Preparation

Salt	*Quality/litre of sterilized distilled water (g)*	*Quality (ml) added to/litre of filtered and sterilized fresh water*
Na NO_3	200.00	1
$NaHPO_4 . 12H_2O$	50.00	1
Na_2 EDTA	3.00	
$FeCl_2 . 6H_2O$	0.24	

Contd........

Salt	*Quality/litre of sterilized distilled water (g)*	*Quality (ml) added to/litre of filtered and sterilized fresh water*
$ZnCl_2$	0.03	1
$MnCl_2 . 4H_2O$	0.27	
$CoCl_2 . 6H_2O$	0.80	
$CuSO_4 . 5H_2O$	0.40	
H_3Bo_3	3.44	
Tris (pH 7.4-7.5)	50.00	1-2

Culture in Carboys

Diatom culture in carboys of 20 litre capacity is maintained to provide a source for culture in large cylinders or for batch culture in larger concrete or fibre glass tanks. Like the test tube, petridish and flask, the carboy is also cleaned by rinsing with fresh water followed by autoclaving and steam cleaning. The carboys containing culture medium are then inoculated with entire contents of flasks containing cells maintained in carboys for five to seven days. When density of cells reaches at 10^6 cells/ml provide vigorous aeration for culture in carboys.

Culture in Large Cylinders

Cone shaped cylinders (100 litres) made of fiber glass are invariably used for cylinder culture. Every cylinder is provided with a bottom out let for employing it, unlike the test tubes, flasks for carboys or cylinders are only steam cleaned. The cultured cells of carboys are transferred to the cylinders with nutrient medium. As in carboys, here, also vigorous aeration is required. These cultures are maintained for two-three days before their transfer to batch culture tanks.

Batch Culture

This involves the mass culture of diatom in outdoor circular tank (500 litre). The bottom of these tanks slopes to a central outlet for emptying, like cylinders. These tanks are also maintained with cultures only during the rearing period of the larvae of rotifers/ *Daphnias*. Outdoor tanks containing filtered fresh water are filled with a fertilizer mix, the composition of which is given as: Ammonium sulphate 100g/m^3, Super phosphate 20g/m^3, Urea 8g/m^3.

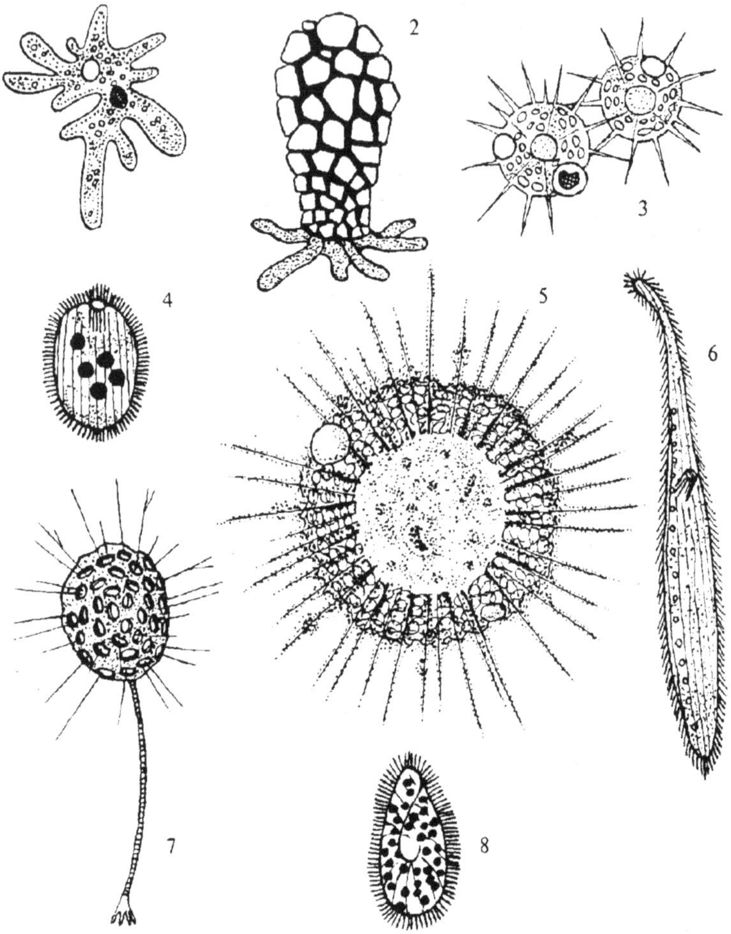

1. **Amoeba proteus; 2. Difflugia pyriformis; 3. Actinophrys sol; 4. Prorodon teres; 5. Actinosphaerium eichhorni; 6. Dileptus anser; 7. Cladrulina elegans; 8. Paramecium bursaria**

The cylinder cultures are transferred to these outdoor tanks and are maintained for one or two days. It would be better if tanks are placed in areas without direct sunlight.

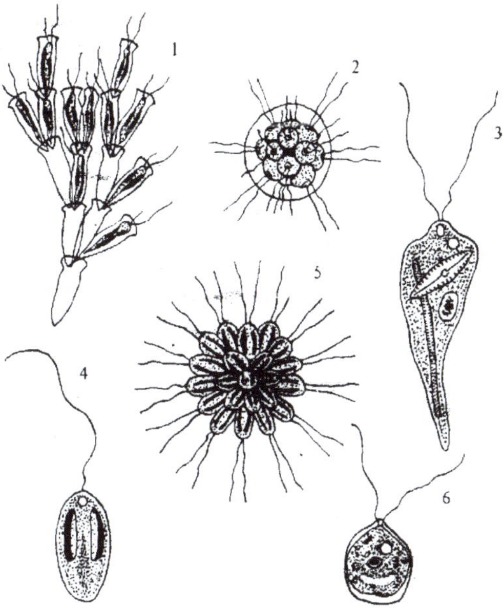

1. Dinobryon sertularia; 2. Pandorina morum; 3. Heteronema nebulosum; 4. Chromulina; 5. Synura uvella; 6. Chalmydomoas

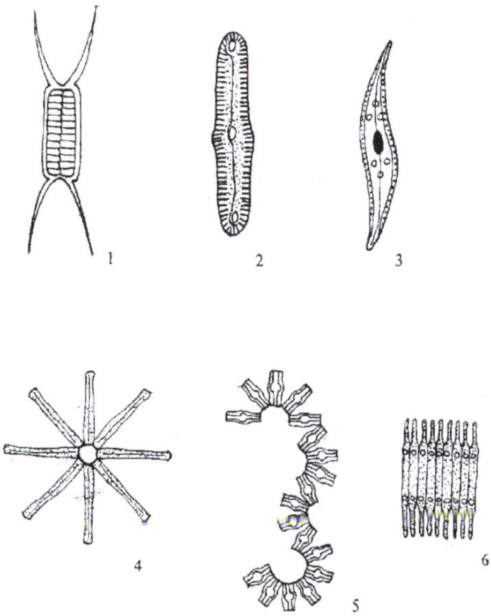

1. Attheya zachriasi; 2. Pinnularia viridis; 3. Pleurosigma acutum; 4. Asterionella gracillima; 5. Tabellaria fenestrata; 6. Fragilaria crotonensis

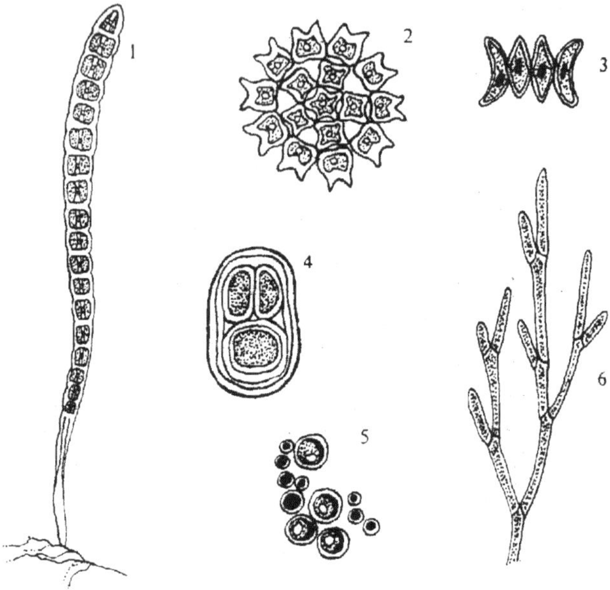

1. Ulothrix zonata; 2. Pediastrum sp., 3. Scenedesmus acutus; 4. Chroococcus giganteus; 5. Protococcus olivaceus; 6. Cladophora sp.

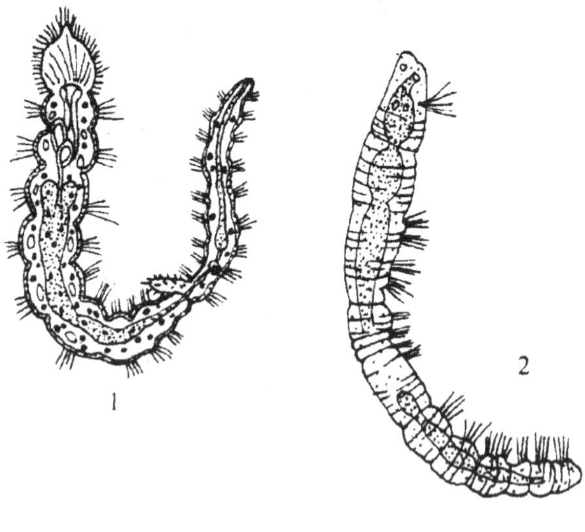

1. Aeolosoma sp.; 2. Chaetogaster limnaei

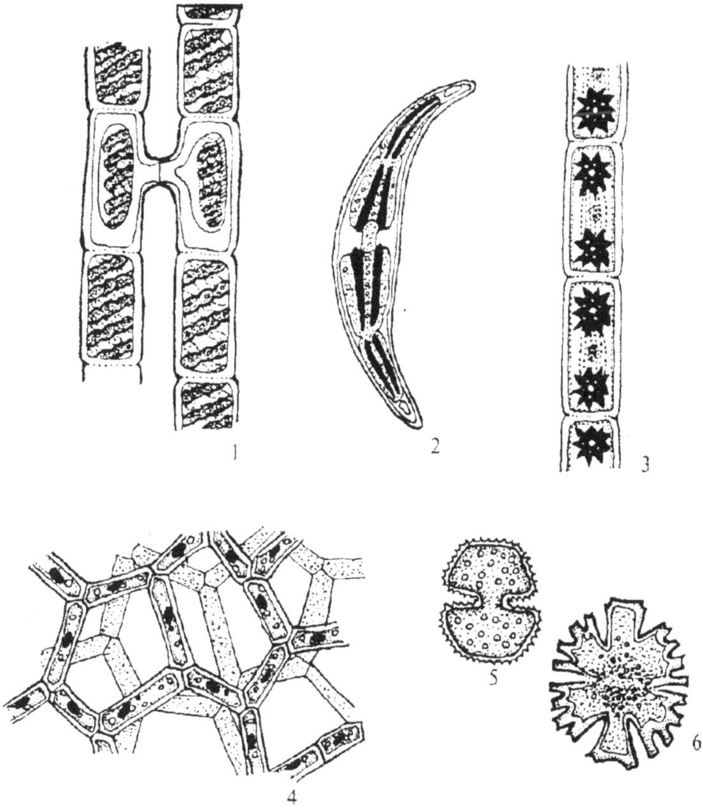

1. Spirogyra sp.; 2. Closterium Ehrenbergi; 3. Zygnema sp.; 4. Hydrodietyon utriculatum; 5. Cosmarium botrytis; 6. Micrasterias apiculata [Figures from O'Conell's (1971)]

Culture of *Artemia*

Artemia are commonly known as sea monkey or brine shrimp, inhabits the salt *pan* waters of more than 200%. There are saffron red coloured organisms swimming cysts of *Artemia* could be collected and stored for a number of years. The cysts are of commercial importance as hatched nauplii serve as an ideal protein rich food source to many fish and prawn larvae in culture systems. For laboratory hatching of *Artemia* cysts, the dried cysts are allowed to float on surface of filtered normal seawater where they hatch in 24-48 hours depending upon the ambient temperature. Eggs can also

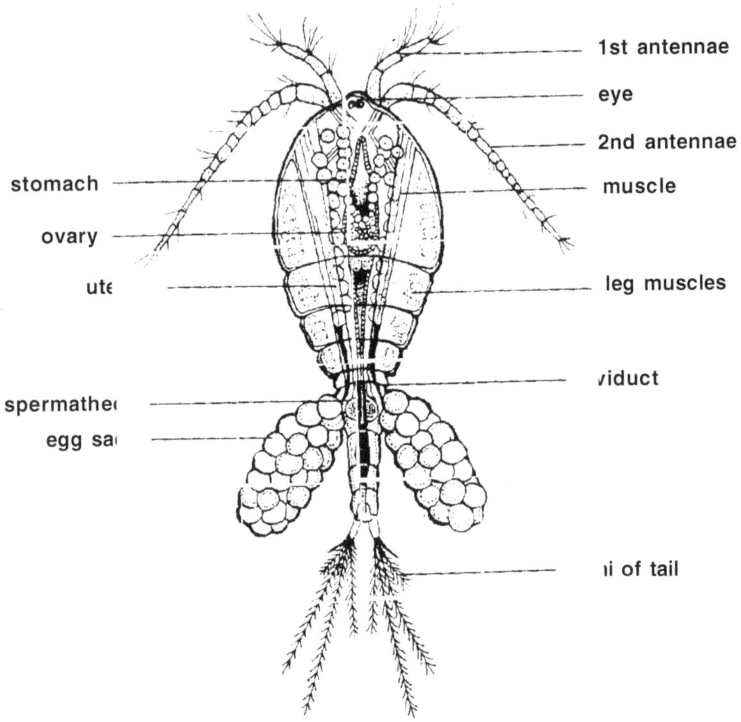

Anatomy of the Copepod *Cyclops albidos*

be incubated in a solution made up of 2 teaspoon of common salt, mixed in 1 litre of fresh water. For mass culture of *Artemia*, about 250 gm of cysts are sprinkled in 100 litre of sea water 30-35% vigorous aeration suitable temperature (26-30°C) may hasten the hatching process. The hatched nauplii could be attracted, using a light source and collected by a very fine meshed net they are either transferred to culture chamber containing fish and prawn larvae as live food or can be stored frozen as rotifers and *Daphnia* for future use. In order to have a continuous supply of eggs, young brine shrimps may be fed from 3rd day onwards with yeast and unicellular algae such as *Isochrysis or Diffusia*.

Culture of *Euglena*

Euglena is found in abundance in pond water, having rich nitrogenous waste matters. For culture take a medium sized battery

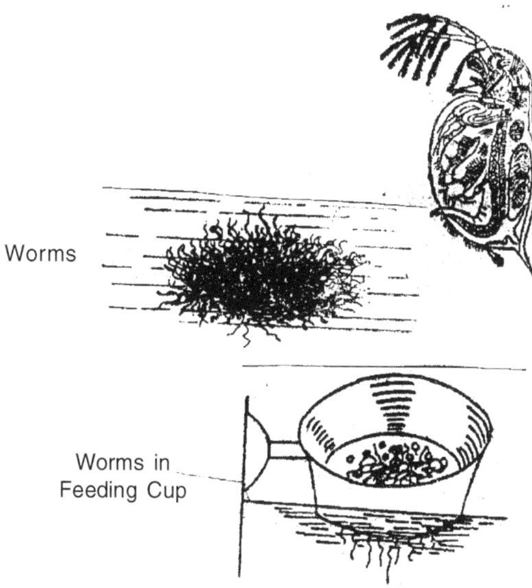

jar and fill it with water and add to it 100 gms of wheat or rice bran and some hay. Keep this solution near the window so that sunlight do not fall directly and leave it for seven days. Now add some pond water containing euglena and watch it under microscope. In two weeks, the water becomes greenish brown and surface of water will be covered with a scum containing large number of *Euglena*.

Culture of *Paramecium*

Paramecium is found in fresh water rich in decaying organic matter. The culture of *Paramecium* is one of the most easy among protozoan as it reproduces very rapidly. To culture *Paramecium* take 15-20 grains of wheat and some hay and boil them in 500ml of water for a few minutes. All are allowed to cool down. Afterwards collect some, and add water having submerged leaves containing *Paramecium*. Add the pond water to culture solution. In a few days *Paramecium* will appear. The best temperature to cultures in is 80-85°F.

Spirulina

Spirulina is a blue-green alga belonging to the class Cyanophyceae, order Nostocales and Family Oscillatoriaceae. It is recorded from water-logged rice fields, ponds, lakes, wastewater channels and brackish-water pools, with alkaline and saline soil substratum being most suitable for its propagation. The research work in India with regard to utilisation has largely been carried out with two species, viz., *Spirulina platensis* and *S. fusiformis. Spirulina* has been known for its protein and beta-carotene apart from a host of vitamins, minerals and pigments in its make-up. It stands out among the algal forms with a protein content of 62.5-71.0%. (on dry weight basis as compared to a maximum of 50% in others). Its protein yield per unit area is the highest among protein yielding crops, with an excellent aminogram comparable to amino acids sequence egg protein patterns. The Proximate composition in percentage is protein 62.5-71.0%, lipids 6-8%, carbohydrates 15-17%, crude fiber 0.02-0.90%, ash 7-9%.

Culture

The commercial design of Spirulina culture system comprises a network of cement raceways with paddle-wheel agitation to prevent alga accumulation at the surface and to ensure uniform distribution of nutrients, with modifications in terms of vertical reactors, tubular reactors, sloping tanks with baffles, etc. The outdoor cultivation parameters are the cement raceways or vats (length-breadth ratio 1.5: 1) preferably with a mid-rib for facilitating water circulation, culture medium depth 15-20 cm, agitation/flow rate 20 cm/sec, temperature 25-35°C, light intensity 20-30 lux, hardness 120 mg/l, pH 9-11, provided with nutrients.

Chapter 8

Breeding of Aquarium Fishes

Introduction

Breeding of Aquarium fishes is a matter of great interest and a highly paying venture to many aquarists who specialize in breeding. As you know that keeping the aquarium in houses, offices, shops etc. is becoming very popular. Nowadays the colourful aquarium fishes are simply not available in abundance in any natural resources because most of them are exotic species. Thus the aquarium fishes command a respectable prize in market and they are sold by professional fish breeders in aquarium or ponds. If we know the technique of breeding the fishes successfully, one can develop it as our hobby and can earn too, thus gets profits from the hobby.

Breeding of aquarium fishes is an altogether a seemingly simple but complex process. We may find one aquarist saying that it is very easy to breed angle fish and he can breed them on order. But at the same time you may find another aquarist, according to whom it is very difficult to breed angle fish and his attempts have ended in failure. The fact is that to breed a species successfully everything has to be right, such as the species or the matched pair the temperature, the light, the quality of water etc should be right. All have to be appropriate and according to the requirement of particular species. It can be said that when an aquarist complains the failure of his attempts to breed a fish, means that he was not able to create the desired conditions in the aquaria. To create these desired conditions, one should not try to get it by trial and error method. Only, rather he should study the actual phenomenon related to breeding and performs the exact ways.

In the forthcoming text, and attempt to outline the general principles of breeding of tropical fishes has been made. The act of reproduction of fishes is not at all similar to other vertebrate animals. Courtship may be evident in the case of fishes but the act of copulation is absent in many fishes owing to this fact the sperm of male fish fertilises the female egg, not in females body but the external environment i.e. water. But in so many fishes viviparity is also common.

Clean the Aquarium before Breeding and make all arrangements

Aquarium Vacuum Cleaners

Despite the best efforts of aquarium filtration, a certain amount of waste excretory matter will accumulate eventually in the aquarium. This is known in aquarium parlance as mulm and is composed principally of the breakdown products of fish excreta and to a lesser degree decomposed food. The simplest way of dealing with this problem is to make sure that your gravel slopes at a sufficient angle to the front of the tank, then with the aid of any one of the several aquarium vacuum cleaners you can siphon up the mulm into the vacuum bag which strains the aquarium water, the dirt being retained in the bag as in a domestic carpet vacuum cleaner. The power source for the siphon can be a simple rubber squeeze-bulb or an air jet supplied by your air pump, both of which will cause a column of water to rise through the cleaning unit. There are also one or two electrically powered models. Most models are effective but the filter bags eventually wear out and have to be replaced, so make sure your dealer stocks spares. Compare the price and quality of these cleaners as there is little to choose between the various models in terms of effectiveness.

Miscellaneous Equipment

Ozonizers

A great deal has been written about the importance of ozone to the fishes' health in suppressing bacterial infection in the aquarium. Unfortunately there have very few scientific studies to demonstrate the claimed advantages and the evidence to date is both conflicting and substantially unquantified. Ozone in the air can be a definite irritant to human beings exposed to levels above a certain threshhold value, and there is no known way of measuring precisely the ozone quantity in salt water where it forms a residue whose side effects may be undesirable. The indiscriminate use of ozone should be avoided except in the case of known bacterial disease, where in order to prevent a build up of contagious bacteria in the water, it may be introduced with the air supply for limited periods at a rate of approximately 1 milligram per hour. Papers presented at the 1977 International Ozone Institute Conference in Paris substantially supported the view that great care should be taken when using ozone. In particular, it should never be introduced where young fry or larvae are being reared, because even very low levels have been found to affect them fatally.

Breeding Traps

For those aquarists who are really serious about their hobby and who intend to breed fishes on a reasonably large scale, this section will be of only limited interest as the establishment of separate tanks for breeding and rearing, perhaps with custom-built traps for individual species, can be the only satisfactory system. Nevertheless, aquarists who are restricted to one or two tanks may find a commercially-produced breeding trap useful to protect the fry of livebearers. There are several models on the market. Some are constructed with a grill, above which the pregnant female is kept. After the fry are born they can swim down through the bars of the grill where the mother fish cannot follow. In certain models the fry find their way out into the tank proper by way of a slot in a V-shaped taper to the trap. Both these types are useful in a community tank if it does not contain other carnivorous fishes and has plenty of plant cover on the surface. In the type of trap which retains the fry in the underneath chamber, the female must be watched carefully to see

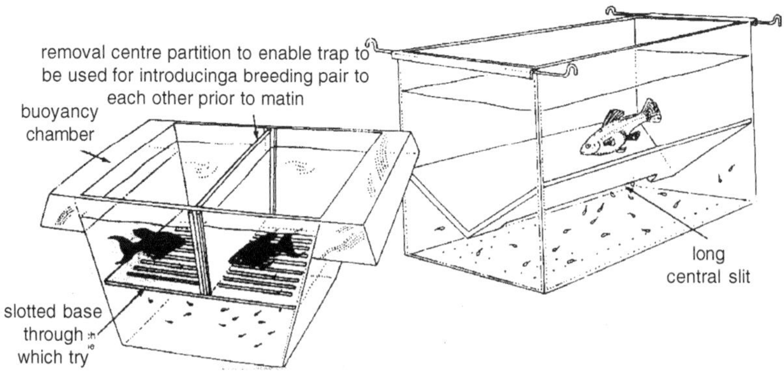

Typical Breeding Traps

when she has finished giving birth, because the young fishes may swim up again and then be devoured by their mother. This trap is often claimed to be useful for protecting newly-laid eggs but traps of this type are too small to have much chance of inducing most egglayers to spawn, and all claims to the contrary should be regarded as suspect, notwithstanding the drawings on the manufacturer's packet. A large breeding trap has another use which has nothing to do with breeding. It can be used to isolate a newly introduced fish for a day or two to give the established inmates, if they are aggressive and territorial, an opportunity to become used to the new fish and accept it. This technique is particularly useful with marines.

Sex Organs of Fishes

The gonads of fishes are of two kinds: ovaries (hard roe) in the female and testes (soft roe) in the male, its fluid also called as milt. The ovaries are pinkish or yellowish in colour, granular in texture, elongated in shape, paired and more or less intimately associated with the kidneys. They lie just behind and below the air-bladder, if one is present. The testes of the male occupy much less the same position in the body as do the ovaries in female, but they are paler in colour and have a creamy rather than granular texture.

Act of Reproduction: Oviparous Species

The act of reproduction is very simple. The germ cells/gametes are primarily fragments of protoplasm containing nuclei that develop into ova in ovaries and spermatozoa in testes. When the germ cells are ripe, the fish betray exuberance, their colours improve and even the most sluggish of them become extremely active. Soon the male begin to chase female. The drive, which continues until the female is stripped of eggs. It is an attempt on the part to move to jump and bump the abdomen of female and so facilitate the release of the eggs. The eggs are released in batches by a special aperture or the excretory, pore and are immediately fertilized by male.

In most fresh water fishes the eggs are adhesive. They attach themselves to rocks and stones, to the leaves and stems of aquatic plants and if the fish are bred in captivity on the glass of aquarium, the number of eggs released varies greatly with the species, much depends on the size of fish. In general, younger fish produce fewer eggs than older fish, and active species, that pay no attention to their offsprings, usually produce many more eggs than those species that protect them.

The unfertilised eggs shed into water are fertilized by the entry of spermatozoa into the egg. Development starts immediately by simple division of egg into two halves. Further, cleavage occurs accompanied by regular growth and blastula is formed. The blastula is converted to gastrula stage by pushing of one side, the centre cavity which forms the alimentary canal. Thus organogeny/ specialization has now begun and these after development proceeds rapidly. Step by step, until the embryo comes into being and later emerges out to form the egg as a larva. The period of incubation may ranges from 12 hours-2 weeks, according to the species and temperature of water.

In general appearance the larva resembles a tadpole. Its most prominent feature being the yolk-sac carried abdominally. At this stage the larva hangs at and rest (occasionally labouring for short distances). Some to the aquatic plants, other to the sides of aquarium, still other at the surface of water. On no account they should be disturbed. No food, need to be given to them; for they exist by absorbing the remains of yolk. This, however, does not last for long, and once the yolk has been absorbed the little fish (post-larva) is free swimming, and strikes out in search of the microscopic organisms, present in all natural waters, on which it feeds. Hence, forward its growth as determined as much from without or from within, and the aquarist who aims to raise a strong and healthy brood of fish must take steps to feed the post larva. Aquarium can be supplied with natural food for very long.

Act of Reproduction: Viviparous Fishes

The great majority of fishes are egg layers, or oviparous species, there is however, an interesting group of fishes known as LIVE-BEARERS or viviparous species those give birth to their young ones. Among aquarium fishes, majority of live bearers belong to family-Poecilidae, a family that is characterized by the males having a modified anal fin which is termed as gonopodium. Most recent researches have shown that true copulation does take place in viviparous fishes. It occurs when male and female, during the course of *nuptial* drive, come close together and for a period that lasts upto about 5 seconds appear to be glued side by side.

It is noticeable that female poecilids have a marked aversion to the males, though some are more approachable than others. From this it has been deduced that the species has not to become extinct the unwilling females have to be attacked by the males when opportunity comes. Eagerness in males of live bearers can be seen to an-extreme when they keep chasing females constantly, even those quite immature, and they even chase the members of their own sex.

The eggs are fertilized while still in the ovisacs in the walls of the ovaries, or within the cavity of ovary itself, and may undergo their development in either position. Embryos are nourished by yolk contained in ova. The oxygen necessary for respiration is supplied by blood of mother. The fish offsprings, more or less replicas of their parents, are born at intervals of few minutes, they pass down the oviduct and enter water usually by tail first. At once they labour to the surface of water, for, when born the air bladder is in a collapsed state and if it is not filled by gulping air at surface, soon after birth, it will remain permanently collapsed. The number of young ones laid vary from a few as 25 to many as 200, depending upon the size and age of female, 40-50 is average at a time in one batch and over a hundred is exceptional. The period of gestation depends upon the temperature. At temperature of about 5°C the period is from 4 to 5 weeks, but a few degrees either way makes a difference. It is to be noted that one impregnation by a male is sufficient to fertilize several successive broods in a female, who delivers the young ones at intervals from 1-3 months according to the species.

Courtship Among Fishes

Courtship, in true sense of word, does not exist and the habit of two kissing gouramis (*Helostoma temminckii*) approaching each other with spasmodic protrusions of lips-suggestive of kissing—has no sexual significance. The lips appear to be adapted for the less romantic. The nearest approach to a definite courtship occurs in a few species in which the male gives an elaborate display before pairing. Even so, it lasts only for the mating period or at best, for one season. The male siamese fighting fish (*Betta-splendens*) is an example of this. This nuptial display is dramatic, with extended fin, mouth gaped, and gills to reveal red membranes to swim around the female, displaying his iridescent colours, which under the stress of sexual excitement, are even more vivid then usual. It is at this period that his pugnacity is at its height, and introduction of second male result in battle royal for favour of female. Many other fishes to show pugnacity at breeding time, among some Cichlids the female is good fighter as male and is as murderous in selection of her mate, as *Betta* spp. is.

Parental Care Among Ornamental Fishes

Comparatively few fishes reveal a sense of parental care. Most eat their eggs as soon as they have been deposited, and the larvae as soon as they have hatched out. Therefore, the aquarists have to take the precaution of removing the brood fish from the aquarium as soon the fish has spawned.

However, exceptions are there, and parental care usually takes the form of the construction of a nest for the reception of eggs. Among the freshwater sunfishes (*Centrarchidae*) the male scoops out a shallow, basin like cavity at the bottom of the aquarium, carefully removes all pebbles and lines it with snout, after the eggs have been deposited. Fish guards them carefully, keeps the nest clear of sediment and remains on guard until the young fish are able to swim freely.

Same type of habit is found among some of the Cichlids, but usually both parents assist in the task of protecting the eggs. The parental care continues even after the young fish are free swimming. It is a rare sight to see a brood swimming in a school escorted by the parent fish, in later stages it shift the young ones to a secured spot, out of any danger.

Many of the anabantids build a nest of bubbles. The male blows bubbles of air and sticky mucus which cling together to form a mushroom-like floating mass of foam on the surface of water. The siamese fighting fish (*Betta*) after her courtship and the surrender of the female turns over her back. Soon he relaxes the pressure and takes up a position below her. She now releases the eggs. As they are heavier than water they float down to the male, who catches them in mouth and carries them to the nest where he remains on guard to protect them from female, whose inclination is towards eating them. The young hatch out in 24-48 hours. They can be seen as small black spots, hanging vertically from the nest. Here, they remain carefully, guarded by male who replaces any then leave the nest before they are old enough to tend for themselves.

In closely related paradise fish (*Macropodus*) the method of spawning is much more the same, but as the eggs are higher than male the female spawns on her back and eggs float in nest without assistance of male. If they fail to reach the nest, female and male collect and put them there.

Arnold's Characin (*Copeina arnoldi*) spawns above the surface of water. The male and female leap together, out of the water and cling together to spawn. When spawning is over the male takes up a position a short distance from nest. Every quarter of an hour male goes below the nest and by violent lashings of his caudal fin, sprinkles the water over eggs. The eggs incubate in a matter of about three days.

Among tropical aquarium fishes, most remarkable form of parental care is practiced by *Megalochromis* and *Tilapia*, who protect their eggs by carrying them in their mouth, mostly the females do so, but in Tilapia this is done by male. The female carries the egg in mouth for 14-18 days, during when female takes no food. Hatching occurs in mouth and after 3-4 days of hatching, only the young ones are allowed to swim out. But if any sign of danger they return back to mouth for protection. Taking a broad and scientific work of parental care in fishes, it can be said that fishes have a sense of possession but something like a protecting father or mother-love is not true. Any adult fish would like to hunt something smaller than itself, irrespective of the thing whether it belongs to its own species or not. Though, it can be seen in most of the animals that they do not hunt their own species.

Line or Selective Breeding

In theory breeding of fish is not difficult but practically many problems are faced and to be solved. It is not enough to mate two fish and rear their offspring to maturity. To make breeding a worthwhile hobby, rather then a commercial venture where aim is to breed and sell more number of fishes in shortest possible time, so, the breeding has to be systematic. Technically called as line breeding. Under systematic cultivation, animals show of great susceptibility to modification and therefore, by selective breeding, desired quality of an animal can be improved in two ways. Firstly individual specimens can sometimes be produced possessing more good qualities that can be found rarely in any individual of original stock. Secondly, from the same stock many variations can sometimes be produced, each characterized by a different perfection. But unfortunately for breeder, nature sometime rather more often tempers with the work of breeding as a result of which a particular quality is improved at the cost of other quality/qualities of the animal. Thus, line breeding necessitates great care and judgement, first to produce of effect, secondly to produce it with the minimum possible sacrifice of other qualities.

Many complications arise when the breeder begins to put theory into practice. He has to take into account the age of the parents, their body vigour, the relative ripeness of the germ cells, the normal time of mating and other factors related to reproduction like environmental condition etc. It should be remembered that as the fishes are kept in aquaria to breed, they are living in artificial surroundings, they acquire artificial habitat and habits and receive artificial food to a great extent. These factors tend to induce acquired physical qualities and peculiarities which are quite distinct to those that are inborn or germinal and which are the secure foundations of success of line of breeding. There are qualities which appear in occasional mutations, but not to be fixed by breeding. In the same way, success can not be expected if the quality is due to habit and habitat, surroundings, and not to be in born of genetical variations. It is assumed that in the development of new strains, nature takes the first step. Finally, the practise of line breeding can be carried to a point that degeneration sets in and in fusion of fresh brood in and attempt to save the strain. It may not be followed by the desired effect, and gives an entirely new turn to reproductive events, and raises fresh complications. Thus, many problems have to be faced by a line breeder. It may happen that new variation or breed may be developed quickly in some cases while in other it may take ages. Demanding infinite patience, most *sedulous* care and a profound study of genetics. Genetics is the science that treats heredity and variation. The laws of Mendel can be scientifically applied to breeding of aquarium fishes.

Selecting Brood Fish

This is very important aspect in relation to selective or line breeding. It signifies that the line breeder should choose the brood fish with great care it possesses all the traits that are desired in off springs. Naturally this is not possible if the breeder is aiming to improve some quality that appears only in one sex, e.g. the sword of sword tail (*Xiphophorus helleri*) or the sail fin of male of *Molliensia latipinna*. In this context, however, the pedigree of female brood fish must be considered. She should be chosen from a strain in which the males are known to have the desired characteristic feature well developed This underlines one of the essentials of line breeding, namely the pedigree is all important, because good fish are always descendants of good parent fish. Finally, it is to be noted that it is unwise to attempt to build up at once a strain possessing all the most desirable qualities. This tends to produce only average good fish and none with the desired qualities to any market extent. The more satisfactory method is to concentrate one or two qualities at a time. This will result-in good fish in the end, even though the fish after few generations will become inferior. Be careful.

Conditioning Brood Fish

Line-breeding involves mating fish that are very closely related. The process exaggerates bad as well as good qualities and if the brood fish are not in first class condition the strain will deteriorate rather than improved. Conditioning the brood fish, therefore, is of paramount importance. Constipation must be avoided and dried food should not be given in any large quantity .The staple diet should be earth worms, varied with *Daphnia*, *Tubifex*, giant-larvae and other live food. The main point to bear in mind is that the fish should be fed to repletion on their natural foods.

Breeding Live Bearers (Viviparous Species)

There are little difficulty in breeding live-bearers. In fact difficulty is to prevent them from breeding when you do not want them to. Three main points to remember are plenty of live food, shallow water at temperature between 75°F-80°F and a thickly planted aquarium with plenty of floating plants. The rest can safely be left to the brood fish. The male should be removed as soon as they have done their task and females as soon as they have delivered the young ones. In the market breeding cages are sold that prevents mother from eating the young ones, but the drawback is that, they allow little room to female especially in the case of larger species. If the exact time of delivery is not known, the female will have to kept in the cage from several days to few weeks, this causes the distress to female and she may die too. Thus breeding cages, are not always suitable and should be avoided. It is good only for the amateurs who count keep a regular watch on gravid female.

Breeding Egg Layers (Oviparous Species)

The egg layers being a vast group, the special requirements of each species can not be described here. However, there are certain general guidelines that have to be followed. The breeding aquarium should be as large as possible, at least 2 feet long, but the water level should not be more than 6-8 inches. One end of the aquarium should be thickly stocked with bushy–foliaged plants, such as *Myriophyllum, Cabomba* etc. The plants need not to be rooted. They should be fielded in bunches. They sink to bottom by toping with stones. They should be so thick that the fish have some difficulty in making their way through them. The aquarium should be kept at a warm place where it will receive the early morning sun. As soon as brood-fish show sign of spawning, they should be placed in breeding aquarium and the temperature slowly raised to by about 5°F. If the conditions are right, the fish usually spawn by the morning after being placed in aquarium. In most cases brood fish should be removed after spawning but those should be left, that protect their eggs/young ones. The eggs should not be disturbed, through infertile eggs attacked by fungus, should be removed by a pipette to prevent spread of disease. The process so far is comparatively easy and the real problems start when the newly hatched become free swimming and the fry has to be raised to maturity. The two essential requirements are heat and plenty of food. The temperature requirement for fry for its proper growth varies from species to species but it is assumed that temperature upto 90°F is conducive. Most important thing is that sudden fluctuation in the temperature should not be allowed to come and young fish can not tolerate these changes. Second important thing for growth of fry is its food, this aspect is being dealt in both in detailed manner in further text.

Guppy (*Pocilia reticulata*)

Family – Cyprinodontidae

Have been named after an aquarist *Robert John Lechmere Guppy*, who had sent a few living pair of this spp. to the British Museum, London in 1860. As Guppies are collected in large number in fresh water bodies they also bear the name *Million fish*. Though the Guppies had their origin in W. Indies, Trinidad they are grown in many countries as *ornamental* and *larvicidal* fish to combat diseaes due to mosquitoes and their larvae. Specially they are used for malaria control.

The female of guppy measures 6 cm and male 3 cm. The former are plain, olive and grayish brown in colour and later decorated with countless colour. According to their colouration the male are named red types, green types, calicopattern, mosauc, fire works, tuxedo, king cobra, albino set. The female are also characterised by wide variation of their caudalfin and somewhat modification in their fins. The male is supplemented with tube-shaped copulatory organ, gonopodium which is modified anal fin. Based on variation in tail the important varieties of Guppy are round tail, top sword tail, bottom sword tail, lyre tail, pin tail, veil tail, double sword tail, delta tail, lace tail, scissors tail, spade tail, flag tail, tuxedo tail, fan tail, king cobra tail, square tail, strange tail are, however, common.

Reproduction: The female attains sexual maturity in 3 months fertilization in internal. The gestation period is 3-4 weeks depending on temperature. The brood female is identified by a dark coloured garvid spot in front of anal fin which is fan-shaped. When the colour of this spot deepens, birth of young ones is eminent. The number of fry per spawning ranges from 20-100. The size of female is a significant feature. They may give less or more young ones and it is also possible that the young ones may be produced in phases. The size of female is a significant feature while selecting breeding stock because bigger female has proportionately larger broods. The life span of Guppy is 3½ years and four generations can be raised in a year.

Breeding: Guppies are bread successfully in a community tank in which different species of ornamental fish live. However, the tank must be densely planted the fry are to be protected from parent which will be injured or engulfed by them, after sometime their release. Female brood guppy can also be transferred to breeding trap or other tank in its original form. The male guppy is pestal in colour with few dark spots or line and these may be changed due to certain mutation.

Condition of the tank: Compared with other species of ornamental fish guppies are very hardy under condition in which no other fish can survive the guppies flourish and breed successfully as they are peaceful fish and mixed with others. The water of the tank should be moderately hard and slightly alkaline and best temperature range for maintaining guppy healthy is 25-27°C. The adult guppies are fed with prepared and fresh live food as well as some vegetation. The fry are fed chiefly with nauplie of brine shrimp or other small live food. Exposing female guppy to tobacco smoke is known to cause sterility. In the case of pregnant guppies deformed fry are produced. Development of spinal cord may also be effected when their diet is without vitamin C and A.

	Morning	Evening
	Tubifix	egg + veg.
	Pellated diet	zooplankton

Family – Characines

The fish which is easiest to breed is *Hyphessobrycon flammens.* The aquarium is filled with 8" of water suitable size of aquarium is 12" × 9" × 9". The plants which are suitable are *Myriophyllum* and *Nitella*. The fishes should be introdued in the evening, immediately after spawning the parents should be removed. The eggs hatch at 24-25°C temperature in 48 hours.

Some species like *Hyphessobrycon callistus* are sensitive to strong light. Therefore, tank should be shaded. *H. innesi* requires soft water and 23°C temperature, *Poecilobrycon* requires a temperature of 26°C, *Megalampiodus megalopterus* requires 4.5 pH, hardness 7-10 and temperature 25°C.

Family – Cichlidae

Cichlids care their youngs with great deligence. They require 25°C temperature. Both sexes develop an ovipositor or breeding tube few days earlier to their spawning. They lay 100-1000 eggs. Eggs hatch during 4 days at 27°C temperature. The orange chromid requires large leaved plants and 27°C temperature. The eggs hatch in 5 days. The Egyptian mouth breeder requires 25°C temperature.

Family – Cyprinidae

Barbs are easy to breed. Aquarium should have fine and densely leaf plant. So they can protect the layed eggs. They require a temperature of 24°C for breeding and incubation period is 24-48 hours. The cherry barbs are best breed in single pairs. *Harleuqueil* is difficult to breed and it requires pH 5.5-6.5 and hardness 2 and temperature 26°C. The scissor tail requires 5 inches water depth and pH 6.6-6.8, temperature 27°C, eggs hatch out 5-6 days.

Family – Anantidae

Their majority are *bubble nesters*. At breeding time they become violent therefore breeding aquarium should be larger and well planted to provide refuge. The water should be 2 inches below from aquarium top and temperature 26-28°C. It is advisable to introduce female few days earlier than that of male. Once spawning is over female should be removed immediately. Fries hatch in 2-3 days. After swimming stage of young ones the male are also removed. Live food is provided to them in 1st week and after this *Artemia nauplii* may be given to them. After 6 weeks they are fed with powdered feed. The Gouramy builds rather untidy and undefined formed nests. Eggs float as they are lighter. Climbing perch perform their breeding nearer to the surface. The eggs float on the surface.

Kissing Gourami

(Native to Siam, the Malay Peninsula, Sumatra, Java) The most characteristic feature of this fish is the thick, negro-like protruding lips. They are furnished with fine teeth. The fish is largely herbivorous, take algae from stones, and grazed tender plants. That is all; for these two fish sometimes appear to kiss each other with as much abandon as a newly- engaged couple, the action must not be concurred in the same way. It is not a prelude to mating. The general colour is light olive green, silvery on the belly. The side is adorned with narrow, horizontal, undulating stripes. There is a dark bar at the root of the caudal fin. The spinouts dorsal fin is marked with a dark lateral band, the spinouts anal fin has a dark margin. The female is paler in colour than the male. In the wild it attains a length of about twelve inches, but aquarium fish specimens do not as a rule exceed four or five inches, and eight inches is exceptional. The species very rarely breeds in captivity. No bubble–nest is built. The pair embrace, and the female scatters about twenty-five to fifty eggs, that slowly rise to the surface and float there. As the brood- fish are greedy eaters of their eggs, the breeding aquarium should be divided by a glass partition and the eggs transferred to the unoccupied compartment as soon as possible. The water should be about twelve inches deep, at a temperature of 82°F to 85°F and with a pH ranges from 7.4 to 7.6. The eggs hatch in about three days. The larvae emerge upside down, and remain in this position until they become free swimming. They thrive well on infusoria, and powdered dried food for the first week or ten days. After that they are ready to take brine shrimps moina and other, microorganism and the like.

Aplocheilus lineatus (Cyprinodontidae Panchax)

This growing up to four inches (10.2 cm). This attractive shaped fish from India and Ceylon is a very hardy species and the largest of genus. The body colour of males is silvery with many rows of bright green and golden spots which extends to fins, which are red. The female is pretty , but simpler in colour with several dark vertical bars. The species should be fed with ample live food especially mosquito larvae. Breed in densely planted aquarium. Healthy strong fish produce the best off springs. So feed them rather more frequently than usual for about a week before mating with scraping raw meat, Daphnia, Shrimp, earth worms and small quantities of dried prepared food. Introduce, female a day in the advance of the male to let her acclimatize to the new surroundings. Eggs dropped amongst the plants by the female are fertilized by the milt ejected from the male. At the completion of spawning, the fish or the plants to which the eggs are attached, should be removed to another tank. The eggs hatch in about 12 days at a temperature of 77°F (25°C), the water level of breeding tank should be lowered to about 6 inches (15.2 cm).

Family – Atherinidae

The breeding tank should be larger, well illuminated with fine leafed plants and mats of Riccia. The eggs hang from plant leaves and they hatch at 7-10 days at a temperature of 28°C.

The fish breeding is rather difficult. The breeding tank should be established one and thickly planted with *Saggiterria*. It should be of 8-10 gallon capacity and should have a fair layer of mulm on the bottom. The female must release eggs and take them into their ventral fins and after it the fish release other eggs. The eggs hatch in about 4 days. The youngs feed upon mulm. The breeding temperature is about 21°C. The water should be slightly alkaline (pH 7.2).

Cultivation of *Infusoria*

Infusoria is the highest sub-division of protozoa, but we may use the term for a minute animal life on which young fish feed, also referred to so plankters. The first feeding should be of *Infusoria*. Plenty of them are found in all mature water. But an aquarium has to be supplemented with *Infusoria* and hence their culture to be maintained separately.

Controversy should not be raised whether which method of culture is best. Generally *Infusoria* grow on vegetable matter. Bruised lettuce leaves, sliced raw potato, spinach, chopped dry hay, uncooked lentils etc. have all been lentils found successful for the purpose. The matter should be reduced to pulp, stirred into water drawn from a pond which has no *Daphnia* and allowed to stand for couple of days. With luck plenty of *Infusoria* may develop, though not guaranteed, except foul smell. Culture of *Infusoria* must be kept fresh and aeration helps a lot. Feeding of *Infusoria* may be done by some decomposing vegetable matter.

Post larvae of fish eat very large quantities of *Infusoria* and this is the secret of raising young fish to maturity is to keep them feeding from dawn to dusk. The drip method of feeding are the simplest and thus best. A jar containing a healthy culture of plankter are raised above the aquarium. A length of ¼ inch of rubber tubing is put into the jar and aquarium as siphon. And the run of culture from jar is regulated to a steady drip by means of pinch cork. Fresh cultures are supplied in the same way as and when required. Other method is to add feed in, aquarium few teaspoon, full of culture every 2-3 hours. Irrespective of way feeding two precautions are necessary, firstly, the culture must be at the same temperature as that of aquarium and secondly, the culture must be strained/filter through a very fine muslin. If these precautions are not taken, high mortality may result.

Live-beares are about a quarter of an inch when born and so take larger food then *Infusoria*. Most other, however, depend on infusoria for 3-4 weeks, care should be taken for young ones of species with small of mouth *(e.g. Colisa lalia*–Dwarf Gourami) they may die while consuming larger *Infusoria* by choking.

Dried Foods for Fry

After first week, bigger food may be given to young fish, some fine dried foods, pea- flour cooked to a paste and squeezed through fine muslin and the yolk of hard boiled egg crumbled between finger and thumb are suitable alternatives. But it should be known that live food is always better than dried foods as the later, when feed shows slow growth of fish. Also uneaten dried food pollutes the aquarium water and it may be removed by using a siphon with sucking by mouth rubber bulb to avoid escape of fry, every evening, carefully.

Screened *Daphnia*

At the stage of four weeks *Daphnia* is the usual, natural food of fry. It is given by placing a good quantity of *Daphnids* in a muslin bag and suspending it in aquarium 3-4 (three to four) times a day. *Cyctops* may enter the aquarium along with *Daphnids,* this should be avoided as Cyclops, though are a good food for fish, is harmful to fry at this stage. Brine shrimp (*Artemia*) is a popular alternative to be screened for. Their culture can be done in artificially created sea water. A small cylindrical worm, *(Anguillada silusiae)*, is also a good food to fry at this stage. It found in earth and roots of the trees. However, its culture can also be done.

Feeding of Growing Fry

Fish that are well fed grow rapidly and soon the fry are able to take food bigger than screened *Daphnia* Microchopped earthworms, *Tubifex,* meshed *Enchytraids* and unscreened *Daphnia,* as well as suitable graded dried foods, may all be given with confidence. Growing fish eat very large quantities of food. And many breeders make it is a rule to stock aquarium so full of suitable food that all the young fish have to do is to open their mouth to get in and out the food when they want. However, the precaution is to siphon out uneaten food should always be kept in mind, to avoid pollution of water.

Sorting Fry

The breeder has to sort the fry periodically. This is very important; for as the fish grow they need more and more space. As more as space are given the better is the growth of fish. The first purpose of sorting is to dispose weaklings, and those that are malformed in finnage and body, which occur in every brood. They make good food for carnivorous species. Fry that are growing well should be removed from aquarium and reared together. If they are allowed to remain with smaller fish they deprive them of food. Fry that grow rapidly, develop into coarse fish that are of no great value. Fry that are weak and delicate should be removed and reared on their own. Many will die before reaching maturity and some breeders therefore, do not bother for them. But this can effect the maturity of other fishes. The best fish, however, develop from among the medium sized fry. These are the ones that should be carefully nursed, given the best attention, the best food, the best room. Thus breeder is most likely to find the qualities that he is breeding for.

Sorting must be regular and *culling* must be rigorous. The experimented line breeders whose aim to produce some particular quality, continue to sort and cull the fry until he is left with only 3-4 pairs of best fish. These are the brood fish for next generation and when they get mature he may mate them together or with original parents (Back-cross), raising the young to maturity again and so on generation to generation. Each generation becomes an improvement on previous one because the off springs inherit their qualities from their parents, which are the best fish according to the breeder. In the end the process to a point where on further improvement can be made. The fish are breeding true to type ones. Any attempt to proceed further may result in strain deterioration, for all domestic animals have a strong tendency to revert to type. This is also true with fishes.

Hybridisation

It is an interesting by less important process of breeding in which mating of two individuals of different species or different genera, is done. Though hybrids are rarely fertile in higher animals, fishes being more primitive, hybrids are often fertile or partly fertile. Although there are records of species hybrids among oviparous fishes, e.g. (*Brachydanio* × *Acquidans*), the viviparous fishes have more scope of experimentation. It is so because of the promiscuous breeding habits of male. In aquarium several hybrids have been produced, e.g.

Lebistes reticulatus	×	*Molliensia latipinna*
Lebistes reticulatus	×	*Xiphophorus helleri*
Limia caudofasciata	×	*Molliensia sphenops*
Limia nigrotasciata	×	*Xiphophorus helleri*
Limia vittata	×	*Lebistes reticulatus*
Platypoecilus maculatus	×	*Xiphophorus helleri*
Platy poecilus variatus	×	*Limia caudofasciata*
Platy poecilus variatus	×	*Xiphophorus helleri*
Lutntana artrizona	×	*Platy poecilus maculatus*

Some of these crosses did not prove successful. Hybridization is not as easy, simple, as to put male and female of different species or genera into aquarium to breed. Hybridization can only be successful if the female is virgin. This means virgin stock intended for breeding must be segregated almost from birth and prevented from coming into contact with any members of the, opposite sex of their genus. This, needless to say, involves, a quite lot of problem. The most common strain species is to experiment with Swordtail (*Xiphophous*) and the platy (*Platypoecilus*). Both species are very variable in colour and beautiful. Line of breeding has already produced several different colour strains which are breeding, true to type.

Theoretically hybridization is not difficult. A male of one species is mated with a female of different species or genera. By back crossing a male offspring with original female parent or a female offspring with original male parent, a desired male or female quality can be strengthened. After that inter-breeding of the off spring will show how far the strain is breeding true to type or if more alluring, thinning is needed practically, though a number of problems are encountered. Care and patience with the presence of an orderly mind can yield the results to the breeder.

Hand Spawning/Stripping

Hybridizing oviparous species is sometimes facilitated by hand spawning or stripping. The ripe female is taken in the wet hands and the eggs squeezed out by a soft pressure on her sides, from the pectoral fins towards the caudal fin. When the eggs have been shed, the water is given a gentle stir (to disturb the eggs on the plants) and then the milt of male is squeezed out in the same way and carried by same wave to fertilise the eggs. With very small fish this operation is impossible without killing the fish. In some cases, however, there is no need to squeeze the female. If she is taken in hand, the moment she sheds a few eggs, and allowed to wriggle on the palm of hand, the action is sufficient to eject the eggs, as many as are ejected normally. Fertilize with male milt and stir with feather.

Methods of Breeding (in brief)

Egglayers

1. For breeding mature gravid brood fishes of both sexes should be selected in separate tank.
2. First wash the tank with strong salt solution.
3. Arrange plants for attachments.
4. They spawn. Different type of set up is required for different fishes.
5. Give moderate light.

For gourami, lillies or broad leave Amazon plants should be preferred. It build nest with (Mucin air bubble) below the leaves eggs are released.

Economics of Ornamental Fish Breeding on Small Scale

Capital Cost				
Glass Tanks				
	48" x 18" x 18"	6 numbers	@ Rs. 600 each	3600.00
	24" x 12" x 12"	12 numbers	@ Rs. 200 each	2400.00
(I)	**Stands**			
	48" x 18" x 18"	6 numbers	@ Rs. 500 each	3000.00
	24" x 12" x 12"	12 numbers	@ Rs. 300 each	3600.00
3.	Cement cisterns of 500 litres	6 numbers	@ 500 each	3000.00
4.	Air compressor	1 number	@ Rs. 2000 each	2000.00
5.	Aeration accessories,plankton net, scoop nets, buckets, mugs etc.			1000.00
	Total			**Rs. 18600.00**
(II)	**Variable cost A**			
1.	Cost of brooders 200 pairs			6000.00
2.	Cost of feed, manure's, chemicals			1000.00
3.	Electricity charges			500.00
4.	Maintenance charges			500.00
	Total			**Rs. 8000.00**
(III)	**Fixed cost B**			
1.	Interest on capital cost	@ 12.5%		2094.00
2.	Interest on working capital	@ 14%		1120.00
3.	Depreciation cost	@ 20%		3360.00
	Total			**Rs. 6574.00**
(IV)	**Total cost (A+B)**			**14574.00**
(V)	**Gross Income**			
	Sale of 1 Lac fish	@ 50% per		50,000.00
(VI)	**Net Income**			**35426.00**

Remove female after egg layering and after hatching (guards the nest) same in the siamase fighter fish.

Angel fish release adhesive eggs and attach them to plants with broad leaves exhibit parental care.

Live bearers

Eggs of fishes are fertilized internally. (Viviparity).

Males are generally smaller and having deep attractive colours or possess copulatory organ gonopodium-a modification of anal fin for the transfer of sperms to female. Well fed gravid of all species release young ones from 10-100 babies at a time. During breeding period parents should be kept separately.

Dense aquatic vegetation give shelter to young ones are released 5-8 weeks after in batches. Provide them small live feed or sieved food material.

After breeding remove parents from tank immediately.

Summary

Fish Name	*Size (cm)*	*Sexual Maturity*	*Fecundity*	*Eggs*	*pH*	*Temp ºC*	*Hatching time*
Zebra *Branchydanio rerio*	4-5	4 months	100-150	Non adhesive	6.6–8.5	15-25	3-5 days
Blue gourami *Colisa fasciatus*	7-9	9 months	700-1200	Nest building Adhesive	6.8–7.8	15-24	3-5 days
Glass fish *Chanda ranga*	6-8	8 months	600-1500	Adhesive eggs	6-8	21-25	2-3 days
Loach *Botia lohachata*	7-9	1 year	3000-3500	Adhesive eggs	7–7.5	10-15	3-4 days
Gold fish *Caraissius auratus*	8-10	8-12 months	1300-2500	Adhesive eggs	7–7.5	15-23	2 days
Angel fish *Pterophyllum sclare*	5-6	6-8 months	100-300	Egg depositor	6.8–7.2	20-26	2 days

Chapter 9

Water Quality

Introduction

Fish obtain their basic necessities from the water in which they live. The most characteristic feature of any aquarium system is therefore the quality of the water it contains. This water must be obtained from some source, pre-treated to make it suitable for the fish, delivered to the fish in sufficient quantities and maintained in good condition. Finally, it must be disposed of.

The water supplied to an aquarium is not pure, but contains dissolved and particulate materials, some are necessary for the well-being of the fish, and others are harmful.

Contamination may occur not only at source or from the animals, but often takes place within the aquarium from the materials used in its construction. The volume of water supplied to an aquarium may at first sight, seem to restrict the number of fish that can be maintained within it. However, it is rarely the quantity of water per unit which limits carrying capacity .The capacity is usually set by the consumption of dissolved oxygen and the accumulation of, toxic metabolic products.

Perhaps the most pronounced and damaging changes to water quality originate with the aquarium inhabitants themselves. In particular, water quality is impaired by the end products of nitrogen metabolism. These include ammonia (either as the gas NH_3, or ammonium ion (NH_4^+) urea, uric acid and other nitrogenous substances including proteins and amino acids. Ammonia, especially, is one of the most harmful substances. The conversion of the more toxic nitrogenous compounds to less toxic organisms residing in water treatment units such as filters. In some aquaria, algae are also used in nitrogen recycling. The process of combating the effects of nitrogenous waste products is facilitated by low stock density, a high water turnover, aeration or oxygenation of the water, frequent cleaning, the removal of faeces and waste food and by the provision of special water treatment facilities.

Dissolved Salts

Fish in freshwater tend to take up water, with a dilution of their body salts and a loss of ions, while fish in sea water tend to loose water with concentration of their constituent salts. In both cases, the fish must maintain the composition of their blood to provide an osmotic pressure equivalent to about 10% (i.e. 10 parts per thousand) salinity.

Freshwater fish cope with body flooding by excreting a highly dilute urine, retaining salts and replacing from their food the small quantities of salts lost. In the sea, fish survive by drinking salty water secreting the excess chloride ions via the gills and excreting isotonic urine.

Salinity is normally defined as the total weight of solid material contained in one kilogram of sea water. Chloride accounts for about 55% of the salinity, and the chlorinity often provides the basis for determining salinity.

Chlorinity can be determined by titration with silver nitrate solution in the presence of an indicator. However, salinity is most commonly determined by measuring the electrical conductivity of the water at a given temperature. A quick determination of salinity may also be made with a hydrometer (the specific gravity of sea water varies with salinity and temperature), or optically by means of a refractometer.

Dissolved Gases

The majority of aquatic organisms need oxygen and must obtain it from the surrounding water. The oxygenation or aeration of the water is therefore, of fundamental importance in any aquarium, especially as the oxygen supply is one of the factors which may ultimately limit the capacity of a particular volume of water for carrying fish.

The oxygen dissolved in water comes from two main sources: the atmosphere and green plants. The actual content is a function of temperature and salinity. Water of low salinity holds more gas in solution at saturation than water of high salinity. Cold water contains more gas at saturation than warm, so that if water is heated, oxygen and other gases will either escape into the atmosphere or the water will become supersaturated. The transfer of oxygen and other gases from the atmosphere to water and vice versa, is dependent upon the ambient pressure, high pressure driving more gas into solution. Thus, a reduction in atmospheric pressure can also result in previously saturated water becoming supersaturated. Oxygen diffuses only slowly through water, and the dissolved oxygen at depth is mainly derived from the circulation of water originally oxygenated at the surface. The oxygen level at depth therefore rarely exceeds that at the surface, though the carrying capacity of the water under pressure is higher.

Oxygen can also be derived from plants as a by-product of photosynthesis, which takes place only in the presence of light (in the dark, plants consume oxygen by respiration). On sunny days, in shallow water, so much oxygen is generated that the surface waters may supersaturate, releasing oxygen into the atmosphere.

Light

Light is perhaps the most regularly periodic of the variables to which fish are exposed, and the one which usually conveys the most detailed information to the fish about its surroundings. Light is essentially directive in its effect, acting upon specific sense organs rather than upon metabolism, but it can trigger physiological responses, an example being the effect of photoperiod upon the endocrine organs.

Photosynthesis generally demands much higher energy levels than vision, and if plants are included in the aquarium, higher light levels are generally required. The photosynthetic efficiency shown by plants depends not only on the light intensity; particular photosynthetic pigments are involved and the wavelength of light is therefore important. For many higher plants at low light intensities, red light is more photosynthetically active than blue, but the difference is not great, and many plants can use a wide band of the spectrum of sunlight.

Where artificial lighting is provided, fluorescent lamps have many advantages. They have a low heat output and long-Iife, though they may initially be more costly. Their diffuse cool light avoids local warming of the water. A wide range of spectral outputs are available, depending on the fluorescent coating of the tube. Warm white tubes have a high yellow/orange content, while other tubes closely replicate the spectrum of daylight, including the ultraviolet component. Some specialist tubes for plant growth have emphasized red and blue components, but can make plants appear pale to the human eye. The spectral quality of a particular tube can be altered, if required, with an acetate filter. The incandescent tungsten filament lamp has a yellow/red base, making it suitable for plant growth and giving a natural appearance to the human eye. In addition, the lamps are relatively cheap to buy, the light is more readily directed and focussed, and more readily dimmed. When dimmed, the spectral composition shifts towards the red end of the spectrum but again the spectral output can be controlled with acetate filters. Diffuse light can be obtained from a array of many low-wattage incandescent lamps behind an plastic screen. High pressure quartz halogen lamps closely resemble daylight and are very efficient, but their initial cost is high. Sodium vapour lamps will provide good algal growth, though their spectral output in no way resembles daylight.

Temperature

Temperature profoundly affects all biological systems. It alters the properties of most biological materials and determines the rate and type of biochemical reactions. Indeed, temperature is perhaps the most potent of all the environmental factors controlling and governing the metabolism of animals.

Water has a high thermal capacity compared to air; that is, it can absorb a large amount of heat energy for a small rise in temperature. It, therefore, provides a thermally stable environment.

In the aquarium, fish are largely denied the use of any behavioural regulation, and the aquarium design and management must compensate for this loss. Sudden changes of temperature in the aquarium should be avoided. Such as thermal shocks are most likely to occur when fish are transferred from tank to tank or when they first arrive in the aquarium complex. A simple rule is to float transfer containers plus fish in their new tanks until the temperature has equilibrated, or alternatively to slowly mix the water in the container with that in the tank over half an hour or more. Increases in temperatures have the most distressing effect since respiration rate and excitability increase while the oxygen-carrying capacity of the water decreases. Such temperature increases in established tanks may result from refrigeration breakdown or thermostat malfunction. The damage caused by faulty thermostats in a heated system can be minimized by employing the minimum wattage heaters required for temperature control or by including a high temperature cutout in the circuit. This could simply be a second thermostat in series with the first, but set to a slightly higher temperature so that it is on all the time during normal operations. Such a thermostat should, however, be serviced regularly to ensure it does not stick on.

Pressure

Fish live at widely different depths both in freshwater and in the sea and are therefore subjected to very different hydrostatic pressures. The effects of greatly increased pressure upon an animal acclimated or physiologically adapted to shallow depths can be profound.

In the aquarium, organisms are rarely exposed to an increased hydrostatic pressure, but they must often live at a reduced pressure compared with their original habitat. In general, fish from as depth as 1200-400 m (pressures of about 120-140 atmospheres) can survive with unimpaired vitality if they are brought to the surface carefully. Creatures from greater depths may suffer more severe stress and show greatly altered behaviour.

Sound and Vibration

It is often forgotten that many fishes are acutely sensitive to sound and other mechanical disturbances of the water. Though the hearing of most species is restricted to low frequencies (below 3 kHz for nearly all fish, and below 1 kHz for most), at these frequency, amplitude, and direction, and many species produce sounds, especially during courtship.

The aquarium is often a very noisy place, with underwater noise levels in aquarium tanks often very much higher than those in the sea or in freshwater. Much of the noise comes from the machinery, pumps and compressors associated with the aquarium, and characteristically contains strong single frequencies in its spectrum. Human footfalls, doors opening and closing, etc., can also be troublesome, and their strongly impulsive nature may evoke startling responses from the aquarium inhabitants. Vibration is transmitted to the water mainly through the floor and tank supports, but also through the water pipes.

References

Hawkins, A.D. (1981). Aquarium Systems. Academic Press, London, pp. 151-168.

Anthony Evans. (1971). Your Book of Aquaria. Fabek and Faber, London, pp. 54-57.

Chapter 10

Common Diseases of Aquarium Fishes

Ornamental fish keeping is a popular hobby in the united states and western Europe and also in India. In U .K. 18% of all households are thought to keep ornamental fish. The classification of ornamental fish keeping can be divided broadly between pond and *"aquarium System"* the latter can be further subdivided into freshwater and marine, and also cold water and tropical. An essential features of successful fish keeping is to maintain water quality parameters within the optimum range for a particular species. A variety of plants and invertebrates may also be kept in aquaria. These, too have specific water quality requirements and it is important to check that these, as well as fish are compatible. The fact that the majority of ornamental fish are imported by air presents the possibility of introducing disease organism into the country. National legislation relating to the prevention of introduction of diseases may make provision for the licensing of imported fish or may restrict areas from which fish may be imported. Where such measures fail, and a disease is introduced. Further legislation may come into effect to prevent the spread of the disease.

Welfare Aspects

The large scale transport of fish in high densities has significant welfare implications. The International Air Transport Association (IATA) has developed a code of practice for the transport of fishes by air and recommendations relate to the types of container, stocking density oxygen etc. One has to follow that those instructions.

Environmental Aspects

Environmental condition outside the optimum range for a species may directly result in the development of specific diseases.

Smaller variations or fish sufficiently to predispose it to disease from other causes. The main environmental factors are:

(1) Abnormal level of dissolved gases–oxygen, nitrogen, hydrogen sulphide, methane etc.
(2) Increased level of metabolites, waste products, ammonia, nitrite, and nitrate.
(3) Additional water quality parameters–pH, hardness conductivity, temperature and suspended solids.
(4) Toxins – heavy metals, chlorine, pesticides, herbicides etc.
(5) Miscellancous factors – Sunburn, trauma, predation etc.

Major Disease

Viruses

Only a small number of viruses are considered at present to be of importance in aquarium fish diseases.

Carp Pox

This results in an epidermal hyperplasia and is found in a number of carp species also. It is thought to be the result of infection with a herpes virus and presents as characteristic discrete candle wax lesions. It is seasonal in incidence, epidemics occurring in spring but disappearing in late summer as the temperature drops. Immunity to the infection is not thought to be strong and there may be recurrence in subsequent years. The disease has also been attributed to avitaminosis and calcium deficiency. For the prevention of disease the ponds, aquaria should be maintained in good sanitary condition and overcrowding should be avoided. This deficiency is also controlled by the addition of chalk to the food (10%).

Spring Viraemia of Carp (SVC)

SVC is the most important disease of ornamental carp. It is caused by *Rhabdovirus carpio*. At infected sites the disease usually develops, as water temperatures begin to rise in spring. Clinical acute disease will occur only when there are temperature above 8°C, and losses may be significant. Clinical signs include lethargy, darkening of the skin, respiratory distress and loss of balance. Abdominal distension (dropsy) and exophthalmia are usually pronounced, and there may be peritonial hemorrhages of the gills and skin. Internal visceral hemorrhages and the presence of serosanguinous fluid within the abdominal cavity are the most frequently reported in postmortem findings. SVC is often in defined as component of the complex of disease known as carp dropsy syndrome. Fish affected by SVC are often also suffering from primary or secondary bacterial septicemia.

Control

This consists mainly of sanitary measures, treatment and solution. Sanitary measures involve strict isolation of the infected school of fishes. Methylene blue is of special value for treatment. The best results are obtained if the fish are given food soaked in a solution of methylene blue for 24 hours. Antibiotics, especially terramycin/chloromycetin/chloroamphenicol have proved especially effective, against the disease. Antibiotics are added to the food for preventive and therapeutic purposes. The dose is 10-20 mg/kg of feed, which is given 2-3 times in early summers. First three days maximum dose according to intensity of infection then taper the dose.

Swim Bladder

This has been described as another component of the carp dropsy syndrome and is thought to be a separate entity to SVC. It usually affects cyprinids ornamental fishes. In their first summer, symptoms being abdominal distension and exophthalmia, in more acute cases, darkening of the skin and loss of balance may be the only symptoms observed prior to death. At post-mortem, pathology is usually restricted to degenerative changes of swim bladder with hemorrhages and necrosis of epithelium.

Treatment consists at present in the systematic administration of food containing methylene blue as follows. Spawners are given 3g of the preparation/kg of dry food for 3 consecutive days. The treatment is continued for 15 days, alternating between medicated and normal food. For under yearling, the medicated food should be given during the season at 20-30 mg per fish in 3 or 4 treatments lasting for ten days, 5 to 8 days apart.

Cichlid Viruses

This virus is thought to be responsible for the acute diseases in South American fish *Apistogramma ramirez*. Clinical symptoms are pale in appearance respiratory distress and hemorrhages of the skin and iris. Mortality is between 40-80%. Treatment by means of antibiotics for 7 days, (ended with tapering dose).

Bacteria

As well as acute systemic and chronic granulomatous diseases, some bacteria may directly invade the epithelium causing damage to the fins, skin ulceration or gill damage. Where as others may infected areas already damaged by other means.

Control: First wash by 1% salt solution for five minutes and then give any antibiotic regularly for seven days.

Acute Systemic Disease

This it can be caused by primary pathogens but is more commonly due to secondary infection of a fish that is weakened by inter current, disease, stress or immuno suppression. The clinical signs reflect the generalized septicemia with loss of appetite. And lethargy, together with erythema and protrusion of the skin and fins. There may be ulceration of the skin. The internal organs may be congested and show hemorrhages, and as cites (dropsy) and expothalmia may occur.

Peracute infections can occur in which there are sudden deaths without prior clinical signs. The organism most commonly associated are the gram-negative rods such as, *Aeromonas, Pseudomonas, Vibrio, Flavo* bacteria, *Yersinia* and *Edwardsiella* species. Specific identification requires the isolation of the infective organisms, usually from the fishes kidney. Some of these organism are associated with specific diseases. *Yersina rudceri* is the causative agent of enteric red mouth (ERM) that can affect gold fish. Where as holes in the head in young catfish results from an infection by *Edwardsiella ictaluri*.

Carp Erythrodermatitis

This disease is caused by the cytophagaceae (slime bacteria). These are generally secondary pathogens or opportunist invaders and the commonest example is *Chondrococcus* species. Various disease condition are attributed to these organisms, but in general they are classified as:

Environmental (bacterial) Gill diseases, Columnaris diseases, Cold water diseases.

Environmental (bacterial) Gill Disease

This is generally a sequel to initial gill damage resulting from environmental factor (such as raised ammonia levels) and is often compounded by parasite infestation the damaged tissue then becomes invaded by myxobacteria, and as the disease progress there is hyperplasia of the gill lamellae which severely compromises gill function.

Columnaris Disease

This is usually associated with stress, nutritional imbalances or other husbandry problems. It is also called cotton wool disease or mouth fungus. The latter being especially common in live bearers. It seldom occurs below 10°C, but explosive epidemic above 18°C. Maintain proper hygeine and balanced diet and healthy environmental condition.

Cold Water Disease (fin rot, peduncle disease)

This requires the presence of predisposing factors, although once established the bacteria will spread to healthy tissue. The lesion generally begin as fin rot progressively spread to involve the base of the tail (the peduncle). Chronic conditions usually seen in cold water from 4-10°C. The causative organism is usually *Cytophaga psychrophilta* and this does not grow above 12°C. So maintain proper environmental condition specially temperature.

Chronic Granulomatus Disease

This disease are generally infectious on set with the formation of granulomates in the tissues. The pathological effect depend on the organ that is affected and granulomate in the skin may ulcerate the surface. The commonest organism involved are *Myxobacterium* spp. which produce chronic to sub-acute disease in all water condition. These diseases are especially associated with overstocking. Transmission is the most usual, although infection through wounds or via ectoparasites may occur. The incubation period is about six weeks and all imported fish are potential carriers. Affected fish are generally restless and may become *emaciates* abdominal swelling together with exopthalmia grey-white nodules can be seen in the internal organs at postmortem.

Flavobacterium spp. have been reported in association with a chronic granulomatus disease in black mollies and other aquarium fish.

Fungi

Various fungi have been associated with diseases in fish, and the symptoms range from superficial vision to those of culminating systemic diseases. Generally, most fungal diseases are secondarily precipitated by poor environmental condition, malnutrition or other primary disease. Once an infection is present with in an aquarium or pond, the level of infective material may increase dramatically.

***Branchiomyces* spp.**

This species produce a disease of gill tissue i.e. gill rot. It is seen especially in cyprinids although, probably most freshwater species are susceptible, the spores attach to the gill surface where they germinate. The hyphae then penetrate the gill tissue causing necrosis which gives a typical marbled appearance.

Ichthyophonus hoferi

This results in a systemic granulomatus disease. Heavy infection can result in a 'sand paper' like effect on the skin. These lesions may become melanised or may rupture leaving small ulcers. Extensive growth can result in abscess formation with the presence of grey, white or yellow nodular lesions in the internal organs. Mortality levels depend on the degree of infection and the organs affected.

***Dermocystidium* spp.**

It can form cysts any where in the body but are commonly seen in the skin or gills. These yellowish-white cysts are usually smooth and can be up to 1 cm in diameter.

Verticillium piscis

It has been associated with granulomata in gold fish, whereas *Exopthiala pisciphila* has been isolated from skin lesions in a variety of marine aquarium fish. *Aureobasidium* spp. have also been isolated from necrotic lesion in the liver of marine species.

Protozoan Parasites

Flagellates

Ichthyobod spp. are found worldwide on fresh water fish, they are obligate parasites with both free swimming and attached form on the host. The disease (slime disease) shows the typical symptoms of ectoparasites infestation. The marked irritation of the skin results in behavioural changes including 'flashing' rubbing against objects and breaking the surface of water. An excess of mucus is produced giving the fish a dull grey colouration and slimy feelings.

Amyloodinium spp. and *Oodinium* spp. (freshwater)

These are found in warm waters and hence are only a problem in tropical aquaria.

Hexamita spp.

They are found in the alimentary tract and have been reported in a variety of species worldwide. Disease is more commonly seen in freshwater species. Infection is probably via oral route and become established in the intestine or caeca. The spread of the disease is by the release of encysted form in the faeces.

Ciliates

Ichthyophthirius multifilis is common worldwide and produces the disease known as white spot or Ich. The infective stage is the tomite which penetrates the epidermis of the skin or gills and develops into the *trophont*.

Tetrahymera costiasis

It occurs in warm waters, 'Guppies' and other live bearers are the most susceptible to 'Tet disease' hence the common name of 'Guppy killer' the parasites may swarm in particular areas of the fish, producing discrete white patches.

Mitraspora cyprini

It causes severe kidney pathology in 'gold fish'

Nematodes

Capillaria spp. are generally thought of as pathogens of aquarium fish, causing gut ulceration and emaciation. Stop feeding to infected fish. It often lie on the side and may die.

Comellanus spp.

These are common parasites of many live bearing tropical fish. They are thin red worms upto 1 cm in length which may be seen protruding from the vent.

Therapy

There are a few particular points with regards to ornamental fish that are worth stressing. The current review of water treatment is likely to lead to legislation which may dramatically affect the availability of many of the common proprietary medications.

Economic constraints, although important, are less of a factor when dealing with ornamental fish. In general, they can be treated as individual since they can be caught easily and then with the aid of anaesthetic agents. Typical treatments of lesions can be performed by bath or dipping or injections given.

Antibiotics

These can be administered via the water by inclusion in the food or by injection. Ideally the choice of antibiotic used should be based on sensitivity testing.

Administration *via* the Water

This is least effective and yet probably the most widely used method for ornamental fish. The wide spread, uncontrolled use of antibiotics in the far East has led to the development of antibiotic resistance. It is advised not to use antibiotics in excess in healthy condition.

Administration *via* the Food

The most common antibiotic used are oxytetracycline, oxycyline, oxolinic acid, amoxycillin and the potentiated sulphonamides. These are generally mixed with food, although manufactured medicated foods are preferable since the availability of the drugs is more uniform.

General Antibacterial Agents

Benzal Iconium: A blend of quaternary ammonium compound and is used as a permanent bath at 0.1-0.5 mg/l. It is useful against, myxobaeterial infection, especially bacterial gill disease.

Chloramine T: It is also useful for *Myxobacterial* infection. It passes an bacterial effect against costia, white spot both protozoans and *Gyrodactylus* spp.

Acriflavin: A fairly popular disinfectant used in koi infection.

Anthelmintics

Medication	*Intestinal Parasite*	*Dosage*	*Method*
Praziquntel	Tapeworms	5-100 mg/kg	Stomach tube in feed
Mebendazole	Tapeworms	25-50 mg/kg	Feed
Piperazine citrate	Nematoels	25 mg/kg	Feed
Leva misole	Nematoaes	10 mg/kg	Via water

Common Ectoparasitic Treatments

Medication	*Ectoparaite*	*Dosage and Administration*
Formalin	Protozoa and flukes	250 mg/l – 30-60 minute dip 15-25 mg/l continuous bath
Malachite green	Fungi and some protozoa	2 mg/l – 30 minute dip 0.1 mg/l permanent bath
Formalin with malachite green	Protozoa	25 mg/l formalin – 60 green minute dip and 0.05 mg/l malachite green dip for one hour.
Potassium permanganate	Protozoa (BGD)	1-5 m/l – 60 minute dip
Copper sulphate	Protozoa (BGD)	0.1 mg/l – 10-15 minute bath

Local Applications

Ornamental fish have the advantages that it may be possible to anaesthetise them and treat lesions individually. Ulcers require through debridement followed by application of antibacterial such as pyridine, iodine mercurochrome or copper sulphate paste or typical antibiotics, ointment. Thus ulcer is packed with a barrier cream, suicidal or a base to reduce the osmotic stress on the fish, antibiotics may be given at the same time.

Prevention of Disease

Good Husbandry and Managements

The manifestation of clinical disease often require predisposing factors, which may reflect on the management of the system, good husbandry methods, frequent observation of fish and regular testing of water quality parameters are of prime importance in the prevention of diseases.

Prophylactic Treatments

A low level of parasites may be another predisposing factor to disease. It is impossible to eliminate these completely and since a closed aquarium or pond favours parasite multiplication, it is wise to treat periodically as a prophylactic measure.

Quarantine

Quarantine perhaps one of the greatest risks of introducing disease to an established system is via new fish or aquatic plants, strict quarantine is essential for a period of at least 4-6 weeks should be considered. So the pathogens show the clinical symptoms and fishes may be discarded. A completely separate quarantine unit with its own filtration system and disinfected equipment is required.

Vaccination

A vaccine has recently become available for the protection of cyprinids against and ulcer disease caused by *Aeromons salmonids.*

Fish Pathogenecity and Development of Vaccine

Evolution has been responsible for various basic biological response in animals. The solitary fish does not need very effective immune system while fishes moving in shoals there are an easy way to transmit pathogen from one fish to another need high immunity. As fishes are having tremendous diversity in structure/ function/genetics etc., there are a great differences in protective mechanism. This enables the individual to survive and maintain its homeostasis and environment which is innately hostile. Diseases by various microorganisms can be detected by clinical symptoms/ histopathologically/clinical biochemistry. The antibody and DNA based probes for pathogen detection and identification could be more accurate, sensitive and fast. ELISA, immune dot assay, Westernblotting, Latex aggultination test, Monoclonal antibodies, PCR and Gene Probes are main diseases diagnostics techniques. The key of DNA base diagnostics is the generation of target pathogen though recombinant DNA technology. Although skin and its extension, gastrointestinal respiratory and urinary tracts represents the first major form of resistance to infection and diseases. Another forms of resistance or adaptive responses develop only in response to presence to foreign organism/objects intrusion by developing strategies for resistance, immediate destruction of pathogen on surface layers by secretion of enzymes or lysozymes or acidic medium, show hypersensitivity or take few hours or days to get rid of pathogen/parasites. There are a few fishes which are genetically hardy, having specific and racial immunity. Some times immunity may be acquired naturally or given by vaccine. The most important is the differences between activation of humoral and cell mediated branch. It should be of immunological memory. Most of the fish vaccines inactivated the pathogen by heat or chemical. Heat inactivation causes extensive denaturation of proteins. Chemical inactivation is more commercialised. Microorganisms can be attenuated so they may lose their ability to cause pathogenecity. Vaccination with toxoid induces antitoxoid antibodies which are capable of binding to toxin and neutralising its effects. Recombinant antigen vaccines, synthetic peptide vaccines, DNA vaccines could be developed. DNA vaccines have remarkable immunological memory. Adjuvants are often used to boost immune response. There is a great scope in fish vaccine to avoid chemical use for treatment. This will certainly enhance the fish production.

References

Chubert, D. Gott Friends. Cure and recognise aquarium Fish Diseases.

Connoy, D.A. and R.L. Herman. Textbook of Fish Diseases

Reichenbach-klinke, H. and E. Elkan. The Principal Diseases of Lower Vertebrates. Diseases of Fishes.

Roberts, Ronald. Fish Pathology.

Sniesko, Janislaus F. and Herbert R. Axelorod. Disease of Fishes (Book 3)

Chapter 11

Urban Aquaculture: A New Horizon to Fisheries

For being ornamental, any object of beauty must be seen in respect of its surrounding and the ornamental pond is not different. It should attract and give pleasure to the beholder, yet it should not force itself upon the eye -rather it should appear to be quite natural and uncontrived; it should blend with and be at peace with its surroundings.

Site of the pool or pond must be away from tree and bushes that can screen the sun. Leaves should not fall into them. Rotting vegetation of all kinds can disturb the ecological balance. The leaves and berries of the willow popular and laburnum are particularly poisonous to fishes. It should also be away from petrol and dust. Accessibility to a water supply is useful, they need occasional topping up in summer.

A lot of literature is available on various aspects of design of tanks and reservoirs for irrigation and power. Not much is available on the design and construction of fish pond especially the ornamental fish pond. Except for difference in sizes and scales much can be drawn upon the existing experience and expertise available for construction of tanks.

Formal ponds may be square, rectangular, circular or oval in shape. Such a pond is really only suitable for formally designed gardens here it will be complemented by straight lines of precision–cut hedges, regimented herbaceous and rose gardens and similar situations. The informal setting of natural outlines with rocky shows irregularity of line. Select rock of largest manageable size. Modern housing estates tend to have small fish gardens and this somewhat restricts the style of garden design. It may be of kidney shape or saucer shape with very gently sloping sides so that amphibians can climb out of the pond with ease. Whatever the shape pond may have but must always be wider than its depth. The largest water surface is required. It should not be less than 30 square feet with perfect compatibility with the size of the garden. There should be at least one shallow area and waterlogged area for the growth of native plants. This area is created by making a ridge in the contour of the line between the deep pond and a shallow area near the side and filling it with soil.

Digging is to be done by labour or mechanical digger. Once the hole is dug, underlay into and over the lip of excavation . Underlay with prevent stones coming up from the bottom and puncturing the lines. For ornamental ponds it is very essential to make provision of shelves for plants. Then put the plants in containers pots, the shelves must be so positioned that the depth of the containers pots as well as the full length of the plant is allowed for. Where plants are to be grown in a permanent position the pond the shelves must contain lips capable of holding a 3-4 inch depth of coverpost in position.

Lay the lines in position. Pond sheeting adapts itself beautifully to the shape. Calculate the liner size by adding twice the maximum pool depth to both length and width of pool. Take care to measure overall pond dimensions i.e. not narrowest point e.g. a furnished pool 10 ft. × 8 ft. × 3 ft. deep will require a liner 16 ft. × 14 ft. A number of lining such as cement, brick stones, alkathene sheets overlaid by earth have been tried so far. They pose their own problems and need to be standardized for a given agrohydrological situation. Seepage is measured in terms of volume/area/time, i.e. cc/cm^2/day. This may be as high as 17 cc/cm^2/day for an unlined pond to 1.5 – 5 cc/cm^2/day for brick lined pond. But all linings can not be used for fish ponds, concrete and bricklined pond do not provide a biologically conducive surface to fish life. Bitumen lining gets dissolved to some extent (in big water bodies) and the chemical dissolved are not desirable for aquatic life. Similarly L.D.P.E. lining though may provide water saving but the earth cover, if put on sloping surface of L.D.P.E., sloughs and slips exposing the lining which is again not good for aquatic life. Polythene sheet is deteriorating in few years, it tends to rot above the water liner where it is exposed to sun and air. Its cost is a serious consideration, choose 500 gauge black polythene sheet as this has a somewhat longer life than clear type. Nylon mesh reinforced P.V.C. is better alternative. This is a tougher type of liner which wears well and have a greater resistance to being punctured. Butyl liners are tough electric welded seam which allow it to create various shape (0.75 mm thick roots and ultra violet resistant). Aqualast liners are black rubber liner, durable, flexible and rot proof. Liner underlay is essential to use an underlay to prevent sharp stones from damaging the pool liner. It is man-made matting, clean and easy to transport and it is also cheaper than sand. Being flexible, plastic liners are not affected by ground movement or temperature fluctuation. Before lining a pond, remove all the sharp stones. Now lay a two inch layer of soft sand over bottom or several old newspaper sheets. Then lay the liner over excavation and place them spread over bricks or weight around the overlap to hold the position because steps shelves are there. Add water as the weight of water pulls the liner down and moulds into the shape of the ponds. Gently pull and ease the liner into position, the fold or any type of crease may be avoided.

There are a number of pond plants with a range of varieties and colours. Before planting, all dead leaves and stems should be removed and trimmed. Clean and disinfect them with 0.1% $KMnO_4$ or 1% lime. Planting is to be done in May and June when they start to grow like water lily, Nymphea of, different colours, *Rannunculus Aponogeton, Vallesnaria Ceratopteres, Cryptocoryne Amazon*, sword plant. Marginal plants like flowering rush, common arrow head (*Sagittaria*), Forget-me-not (*Myositis*) must always be trimmed and thin, the plantation otherwise they will be overcrowded.

Fountain operating from centre or showering from one side of a beautifully designed status certainly increase the movement of water which will be advantageous for fishes and other occupants. One can perforate the pipeline with definite intervals. To lift the water, a pump is needed. Always take slightly more powerful pump than required for the pond.

A filter is essential component for pond's hygiene. The physical removal of suspended particles from water through the process of straining the discolouration of water is done by mechanical filtration. In biological filtration toxic wastes from fish decaying plants and decomposing uneaten food are broken down into harmless products that are subsequently absorbed by plants. It also involves the breakdown and removal of nutrients and wastes products from pond water which, if allowed to accumulate will give rise to algae growth with green water. This process is performed by millions of nitrifying bacteria which develop and live on surface of filter medium.

The bacteria rely on a continuous supply of water to provide them with essential food and oxygen necessary for survival. If water supply is cut off for more than four hours out of 24, bacteria will start dying. The filtered water returning to pond is now rich in nitrates, an essential food source for all pond plants. The splashing water falling back into pond also helps to induce oxygen into pond again improving the environment for all aquatic inhabitants. In pools biological foam filter provides mechanical and biological filtration.

The shape of the pool is such that some areas would be left unfiltered if there is only one outlet. This is where bottom drains and multilevel outlets incorporating gravity fed systems come into their own. All pipework needs to be stiff in position before the liner is installed, ideally bedded in concrete so that the subsequent weight of water above will not cause it to buckle. Bottom drains should be spaced according to the size and shape of pond. The aim is to remove the sunken debris from the pool floor; so it makes sense to incorporate lower areas for the bottom drains to sit in.

These lower areas can take the shape of sunken wells, perhaps a V-shaped profile throughout the whole pool floor. Debris will then be naturally drawn through the drains and forced into the filter.

Multilevel outlets work on the same principle as bottom drains though they have two advantages; (a) The position of connecting pipework is on the pool side approximately 12" below the surface. At this point there is less water pressure on liner connector than with a bottom drains and therefore, less chance of a leak. Also, multilevel outlet allow the use of a surface skimmer, so that the debris is collected not just from the floor of the pool, but from the top as well.

The position of inlets for the returning filtered water is as important as the position of the outlets, since this water will be returned under pressure from the pump sited in the filter chamber, via U.V.C.

It can, therefore, be directed so that it circulates throughout the whole pool, creating currents which naturally sweep debris towards the bottom drains. If these inlets are incorporated in a venture, they will provide added advantage of aeration.

Now stock the pond after two or three weeks until plants have settled down and filter system should operate properly. Remove molluscs and other undesirable organisms from pond and introduce 1 inch fish/10 square inches (excluding tail) (1" fish / 10^2" area) in pond. Calculate the number of fish according to the area of pond. Feed them thrice a day with changed menu of artificial and natural live food. It is compulsory to note that artificial food when supplemented with 10-12% marine weeds gives better results and sharp and bright colouration. Check the fishes monthly. Fishes may be koi, goldfish, guppy, gourami or any fish of your choice.

The use of electric water pumps has greatly amplified the movement of water in pond. It is no longer necessary to lay water supply pipes to fountain and water falls nor it is essential by proper to carry the surplus overflow of water of some drainage joint. Once the pool pond has been established, it should take only a small amount of enjoyable maintenance to keep it looking good.

One may construct his own pond, person having knowledge and cultivation practices of ornamental fishes may earn lots of money through urban aquaculture, he may give consultancy for beautification of the cities, picnic spots, five star hotels, Govt Offices, houses and may also supply fishes to desirable persons.

Be cautious about:

1. The large *Dytiscus* Diving beetle and its larvae can take small fish.
2. Grass snake like nothing better than swimming after a goldfish.
3. Both hedgehogs and badgers have occasionally been found guilty of pond predation.
4. During spring, many members of carp family become quite predatory and think nothing of gobbling up a few medium sized goldfish.
5. Put a net over your pond to avoid birds and frog entry and predation also.

The shimmer of water, the tranquil beauty of lilies blooming and the swirl of rising fish are only a few delights of having a water garden. Among many virtues, the water garden has two qualities that are outstanding. It creates a focal point that so many gardens lack and it offers the gift of relaxation. A garden pond can be as simple or elaborate as you wish. I personally think a garden pond should have a time for everything. Plants, moving water as in a waterfall or fountain and of course fish. Flourishing plants and thriving fish are the aim and reward of a well managed pond. Our garden catalogue is hereby to help you construct and maintain a garden that will hopefully give you and your family pleasure for years to come.

Calendar

A month by month guide for all your pond requirements.

January/February

These months are the coldest months of the year and the quietest time in your pond. You will not be feeding your fish at this time which gives them chance to burn off any excess fat. It is also the time of year to keep a pond heater in your pool.

The heater (use 1 per 50 sq. feet) will keep a small area free from any ice thereby allowing important gaseous exchange.

To prevent the build up of these in your pond we recommend the pool heater.

March

With the arrival of warmer weather your water garden will now start coming back to life.

Bring your pump out of winter storage, check impeller and bearings and switch back on.

The restart your filter introduce a good starter culture and when remaintained switch on your ultra violet clarifier after changing your tube and cleaning the quartz sleeve.

Remove your pond heater but keep it handy in case of sudden frosty periods.

Fish will become more active and can be fed sparingly on a good wheatgerm based food.

Observe your fish closely after their "Winter Hibernation" as they are at their weakest after the winter fast and a prophylactic treatment is advised.

Now is the time to divide any plants from the previous season and repot ready for summer.

April

Temperature should now have risen to a point where fish are feeding regularly. Use a good wheatgrem based food to begin with. Feed sparingly to begin with so not to overload your filter system or the fishes' metabolism early in the season.

Monitor the water conditions for Ammonia pH Nitrate.

May

Time for pond planting. It is always best to use pond soil in hessian or basket liners. If you don't have a U.V. sterilizer it might be time to use an algicide.

May is a good time to introduce any new fish. If you have the facilities quarantine them before introducing them to your pond. If not, keep an eye on them for the first few weeks.

Remember stress is the biggest danger to a new fish.

Take care when feeding make sure all food is consumed quickly. To keep all the food in one place to be able to monitor this, we recommend the use of a the Bludgon Feeding ring. You should have already purchased a nitrate test kit to monitor the levels and if we have a hot summer, you will need to check the oxygen levels in the pond.

Use the tetra or local oxygen test kit.

June

Increased feeding can cause higher nitrate levels. If your nitrate levels are rising it may be necessary to carry out several partial water changes to reduce them. Remember to use a dechlorator when adding fresh tap water to your pond.

July

Increase feeding now. If the water temperature is high fish use enough food for growth as part of the diet.

August

Higher temperatures means increased feeding and the possibility of higher Nitrate levels. Make sure you monitor your pond weekly.

Hot weather can reduce water level by 2″ a week and should be replenished as required.

September - onwards

It is a good idea to treat your fish as a preventive measure before winter.

Continue feeding as long as water temperatures permit. Do not leave any surplus food in the pond at this time.

Remove any dead leaves from your pond plants and use a pool cover net to prevent falling leaves from entering your pond. Floating plants should be moved indoors.

Once the fish have ceased to feed the pond can be prepared for winter. Remove the pump, U/V and filter and clean and store for winter. If you choose to leave your pump in the pool over winter it is a good idea to run for 1 hour each week to prevent any silt build up. Give the base of the pool a good clean out. Get rid of any decaying matter which may rot over winter and bring your pond heater out of storage just in case.

The Siting

Pools enjoy sunlight. Most aquatic plants especially lilies need as much light as possible otherwise they are far less free flowering.

Site the pool away from trees and bushes that can screen the sun. Pools dislike leaves falling into them. Rotting vegetation of all kinds can disturb the ecological balance. The leaves and berries of the willow, popular and laburnum are particularly poisonous to fish.

Don't site the pool near to road where there could be pollution from petrol and dust. A windbreak such as a hedge or rock garden on the north side to protect the pool from cold northern winds is a good idea.

Accessibility to a water supply is useful. Pools occasionally need topping up in summer.

So bearing these points in mind, stand back and examine from your house, your patio or sitting area, where you could appreciate the beauty of a water garden. Will you add a waterfall? A fountain? Ornaments? Garden or Underwater Lighting?

Where will you site an external biological filter, essential for pool hygiene? Consider the accessibility of the electricity supply for the pump and lights. Next consider the shape. For this it's best to lay a rope on the ground to get the proportions right. Will you have a natural or formal pool?

Consider the size, bearing in mind that pools should be a minimum of 30 square feet to create the balanced conditions for plants and fish. In a large pool there is less temperature variation and more oxygen available for the fish.

If you are considering a waterfall the soil from the excavation can form the basis of the rockery.

Installation of Pool Liners

A pool liner has the advantage of flexibility of size and shape. It's up to you.

A liner is supplied in a square or a rectangular sheet. So for ease of construction and less wastage, design and shape with sweeping curves rather than numerous narrow inlets and necks.

An important consideration is depth. Under 15″ deep there is a danger of it freezing solid in winter and being too hot in summer. Having said that, with the exception of specialised Koi ponds, there is no practical reason for a pool to be over 30″ deep. A good average is 18″-24″. Creating a ledge 9″ below the water surface and 9″ wide provides a home for marginal plants. Lilies and other deep water aquatic plants can be planted in the deep area.

Construction is simple. Excavate the pool insuring the walls are at an angle of 15°-20° (3″ inwards for every 9″ depth). Remove any stones or sharp protuberances.

Strips of Liner Underlay should be laid in the excavation to provide cushioning for the liner.

Next drape the liner loosely into the pool anchoring the edges with stones or bricks but ensuring that the liner is sufficiently loose to avoid stretching and then start running water onto the liner. The weight of water will bed the liner snugly into your sculptured hole. Once the pool is filled trim away the surplus linter leaving a 6″ flap timer around.

Now for the finishing touches. The perimeter of the pool may be edged with slabs or paving stones that overhang by 2″ to hide the liner. To enhance the aesthetic appeal even more the liner may be taken up behind a layer of stones or rocks with paved edging laid on

Installation of Pool Liners

top. Thus, when the water level falls due to evaporation the stone or rocks become visible and the liner remains concealed beneath the water level.

How to Calculate your Liner size

Calculate the liner size by adding twice the maximum pool depth to both length and width of pool. Take care to measure overall pond dimensions i.e. not narrowest points. e.g. a finished pool 10 ft × 8 ft × 3 ft deep will require a liner 16 ft × 14 ft.

Butyl Liners

Long established liner. It is a black rubber sheet and carries a 20 year guarantee. Its thickness is 0.75 mm and is ultra violet resistant. Root resistant to DIN 4062. Cold resistant to DIN 53351. Tensile strength to DIN 53455. Welded seam durability. Free of regenerated materials including calcium and other metals. Available in any size. (The trade numbers are given above).

Aqualast Liners

Aqualast is a black rubber liner which has a 20 year guarantee but a life expectancy of 50 years. 0.40″ (1 mm) thick. Durable flexible, rot-proof. Waste material incinerates safely.

Liner Underlay

An absolute must under any liner. It is essential to use an underlay to prevent sharp stones from damaging your pool inter. An inorganic man-made matting so it will not rot. It is clean an easy to transport and cheaper than sand.

Poolglaze

For treating new concrete. Neutralizes lime in cement and reduces porosity. 1 kg will treat 80 sq ft of cement work.

Poolseal

For old and leaking concrete pools. A liquid plastic available in black, blue or stone. Average 2 coats at 10 sq. ft. per litre more on porous or rough surfaces.

Waterproofing Powder

Essential to add to concrete mix for new pools. Use 1 kg per cwt for cement, 2 kg per cwt for rending coat. 1 kg is sufficient for 15 sq ft of pool surface.

Poolseal Primer

For old and leaking concrete pools. To be applied before Poolseal.

Pool Paint

A new water based formula with superior coverage to other brands. 1 litre per 6m. No primer is required. Now in clear, black, stone and blue in conveniently 3 litre pots.

G4 Pond Seal

G4 Pond Seal is a moisture cured polyurethane which forms a non-porous seal on concrete. Because G4 is moisture cured it is not inhibited from curing (drying) by small amounts of moisture in the substrate. However the substrate should be as dry as possible and have some porosity to enable maximum penetration and mechanical key. G4 can be applied in poor weather conditions, high humidity and at temperature down to 0°C (32°F). Seals out lime. G4 can also be used for repairing cracks in concrete ponds and patching glass fibre pools.

Hartel 225 Resin is the new material developed by Lotus (U.K.) for these 3 self contained fountains. They are virtually unbreakable, totally weatherproof and accurately mimics real stone.

How to Construct a Wildlife Pond

The pond should have the shape of a saucer with very gently sloping sides to that amphibians can climb in and out of the pool with ease. There should be at least one shallow area and a waterlogged marshy areas for the growth of native bog plants. This area is created by making a ridge in the contour of the liner between the deep water and a shallow area near the side and filling it with soil. The pond should have a depth of 18-24″ somewhere in the centre to prevent it from freezing solid in the water.

Once the hole is dug, lay underlay into and over the lip of the excavation. Underlay will prevent stones coming up from the bottom and puncturing the liner.

Next lay the liner in position. Pond sheeting adapts itself beautifully to the shape described.

Then another layer of Underlay is placed on the top to further protect the liner, to give, the soil something to bed into and to allow

A rugged water feature with a choice of fountain or waterfall or both. No hole to dig, set up and working in minutes

Height 13" base 42" × 44" finished granite

An easy to install fountain, just place on the ground and decorate with stones if you wish

Height 11", base diameter 30"

A wall mounted ornament held by just two strong screws which are supplied

Height 30½" width, 18¾" depth of bowl from wall 12" Sandstone finish.

you to place pebbles and stones in the pool without fear of puncture. Then a 4″ layer of sandy subsoil is laid on top.

To avoid a mudbath the pool should be filled very slowly. The best way is to place the hosepipe on top of a piece of underlay and let the water trickle over it. This way the soil will remain undisturbed. The plants can now be rooting straight into the soil. Native plants are readily available from established water garden centres.

Bibliography

1. Brett, J.R. (1970) Temperature (3.3 Animals, 3.32 Fishes) In Marine ecology a comprehensive integrated treatise on life in oceans and coastal waters. Vol.I, Part 1, O. Kinne (ed.) Wiley, London 515-560.
2. Brewer P.G. (1975) Minor elements in sea water. In Chemical Oceanography,Vol.1, 2nd ed. J.P. Riley and G. Sikrrow, Academic Press, London.
3. Brewwer, P.G. and J.C. Goldman (1976) Alkalinity changes generated by phytoplankton growth. Limnol. Oceanogr. 21: 108-117.
4. Chave K.E. and Suess (1967) Suspended minerals in seawater. Trans. N.Y. Academic Press. 29: 991-1000.
5. Chave, K.E. and E. Suess (1970) Calcium carbonate saturation in seawater, effects of dissolved organic matter. Limnol. Oceanogr. 15: 633-637.
6. Clark, J.R. and R.L. Clark (ed.) (1964). Seawater systems for experimental aquariums: a collection of papers. Res. Rpt. 63. Bur sport fish, wildl, Washington, DC 192.
7. Degobbis, D (1973) On the storage of seawater samples for ammonia determination. Limnol. Oceanogr. 18: 146-150.
8. Evans, A. (1971) Your book of Aquaria, Faber & Faber, 3 Queen Square, London.
9. Evan, D.H. (1973) Sodium uptake by sailfin mollie *Poecilia laptipinna*, Kinetic analysis of carrier system present in both freshwater acclimated and seawater acclimated individuals. Comp. Biochem. Physiol. 96B: 111-115.
10. Frank W. Orme (1977) Fancy Gold Fish Culture, Academic Press, London.

11. Fry, F.E.J. (1971) The effect of environmental factors on physiology of fish. In Fish Physiology (W.S. Hoar and D.J. Randall) VI: 1-98, Acadmic Press, New York.

12. Gallagher, J.L. (1967) An economical cooling system for aquaria. Amer. Biol. Teach. 29: 335-536.

13. Garnaud, J. (1977) Construction de l' aquarium public Bulletin de l' Institute Oceanographique 72 No.1433.

14. Goldizen, V.C. (1970) Laboratory Culture of Marine Organisms in Food-drugs from the sea Proc. Mar. Tech. Soc. (1969), H.W. Youngken, Jr. (ed.) Mar. Tech. Soc., Washington DC, 113-117.

15. Harvey, H.W. (1941) On changes taking place in seawater during storage. J. Mar. Biol. Assoc. U.K. 25: 225-233.

16. Hawkins, A.D. (1981) Aquarium System, Academic Press, New York.

17. James, W. Atz (1971) Aquarium Fishes, Pelham Books, London.

18. Jones, A.C. (1970) Chlorella for rearing of marine fish larvae. FAO Fish Cult. Bull. 2: 3.

19. Kawamura, S. (1975a) Design and operation of high rate filters. Part 1. J. Amer. Wat. Wks. Assoc. 67: 653-663.

20. Kils, U. (1976-1977) The salinity effect on aeration in mariculture, Meersforsch 25: 201-203.

21. Kinman, R.N. (1972) Ozone in water disfection. In ozone in water and wastewater, F.L. Evans (eds.) Ann. Arbor. Sci. Ann. Arbor M1: 123-143.

22. Kinne, O. (1976) Cultivation of marine organisms water quality management and technology. In Marine Ecology: a comprehensive, integrated treatise of life in oceans and coastal waters. Vol.III, Part 1. O Kinne (ed.) Wiley, London. 19-300.

23. La Que F.L. (1973) Marine Corrosion, J. Wiley & Sons, New York.

24. Mangum, D.C. and W.F. Millhenny (1975) Control of marine fouling in intake systems - a comparison of ozone and chlorine. In Aquatic application of ozone. W.J. Blogoslawske and R.G. Rice (eds.) Internat. Ozone Inst. Cleveland. 138-153.

25. Meijers, A.P. (1977) Quality aspects of ozoniation Wat. Res. 11: 647-652.

26. Menzel, D.W. (1966) Bubbling of seawater and production of organic particles: a re-valuation Deep Sea Res. 13: 963-966.

27. Nagy, R. (1964) Application and Measurement of Ultraviolet radiation. Amer. Indus. Hyg. Assoc. J. 25: 274-281.

28. Nebel, C.R.D. Gottschling, P.C. Unangst H.J., O'Neill and G.V., Zintel (1976) Ozone provides alternatives for secondary effluent disinfection. I. Wat. Sew. Wks. 123(4): 76-78.

29. Neyfert, A.V. (1969) A refrigeration unit for aquariums with running water. Hydrobiol. J. 5: 68-71.

30. O'Conell's, R.F. (1977) The freshwater Aquarium. Arco Publishing Company Inc., 219, Park Avenue South New York -10003.

31. O'Melia, C.R. and W. Stumm (1967) Theory of water filteration. J. Amer. Wat. Wks. Assoc. 59: 1393-1412.

32. O'Neilla, R.V. (1968) A simple marine aquarium system. Turtox News 46: 166-168.

33. Pickering, A.D. (1981) Stress and Fish, Academic Press, London.

34. Pytkowicz, R.M. and E. Atlas (1975) Buffer intensity of seawater. Limnol. Oceanogr. 20: 222-229.

35. Riley, J.P. (1973) Analytical Chemistry of Seawater in Chemical Oceanography, J.P. Riley and G. Skirron. Academic Press, London.

36. Reginald Dutta (1971) Beginner's Guide to Tropical Fish, Pelham Book Ltd., 44, Bedford Square, London, WC 11350U.

37. Schmidt-Nielsen, K. (1975) Animal Physiology: Adaptation and environment Cambridge Uni. Press, Cambridge.

38. Scott, K.R. (1972) Comparison of efficiency of various aeration devices for oxygenation of water in aquaria. J. Fish. Res. Board Can. 29: 1641-1643.

39. Spotte, S. (1974) Aquarium techniques: Closed System marine aquariums. In Experimental marine biology. R.N. Mariscal (ed.) Academic Press, New York.

40. Sprague, J.B. (1966) Filteration of seawater for marine biological laboratories Mss. Rpt. Ser. (Biol) No.851, Fish Res. Board Can.

41. Suzuki, J.K., Hukushima and S. Suzuki (1978) Effect of ozone treatment upon biodegradability of water soluble polymers. Environ. Sci. Tech. 12: 1180-1183.

42. Sverdrup, H.U., Johnson, M.W. and Fleming, R.H. (1972) The Oceans. Prentice Hall Inc., Engleurood Chffs, New York.

43. Wedemeyer, G.A., Meyer Fres P. and Smith Lynn Wood (1999) Environmental Stress and Fish Diseases Narendra Publishing House, 1417, Krishan Dutt Street, Maliwara, New Delhi.

44. Weyl, P.K. (1970) Oceanography: An introduction to Maine Environment. Wiley, New York.

45. William, R.J. (1963) Biochemical Individuality. John Wiley and Sons Inc. New York.

46. Wilson, T.R.S. (1975) Salinity and the major elements of seawater. In chemical oceanography Vol.1, 2nd ed. J.P. Riley and G. Skirrrow eds. Academic Press, London.

47. Zelac, R.E. and H.L. Cromroy, W.E. Bloch, B.G. Dunavant and H.A. Bevis (1971) Inhaled ozone as a mutagen. Effect on frequency of chromosome aberration observed in chinese irradiated hamstress Enviorn. Res. 4: 325-342.